International Archives of the History of Ideas
Archives internationales d'histoire des idées

Founding Editors
Paul Dibon
Jeremy Popkin

Volume 251

Honorary Editor
Sarah Hutton, Department of Philosophy, University of York, York, UK

Editor-in-Chief
Guido Giglioni, University of Macerata, Macerata, Italy

Associate Editor
John Christian Laursen, University of California, Riverside, CA, USA

Editorial Board Members
Jean-Robert Armogathe, École Pratique des Hautes Études, Paris, France
Stephen Clucas, Birkbeck, University of London, London, UK
Peter Harrison, The University of Queensland, Brisbane, Australia
John Henry, Science Studies Unit, University of Edinburgh, Edinburgh, UK
Jose R. Maia Neto, University of Belo Horizonte, Belo Horizonte,
Minas Gerais, Brazil
Martin Mulsow, Universität Erfurt, Gotha, Germany
Gianni Paganini, University of Eastern Piedmont, Vercelli, Italy
John Robertson, Clare College, Cambridge, UK
Javier Fernández Sebastian, Universidad del País Vasco, Bilbao, Vizcaya, Spain
Ann Thomson, European University Institute (EUI), Florence, Italy
Theo Verbeek, Universiteit Utrecht, Utrecht, The Netherlands
Koen Vermeir, Paris Diderot University, Paris, France

International Archives of the History of Ideas/Archives internationales d'histoire des idées is a series which publishes scholarly works on the history of ideas in the widest sense of the word. It covers history of philosophy, science, political and religious thought and other areas in the domain of intellectual history. The chronological scope of the series extends from the Renaissance to the Post-Enlightenment. Founded in 1963 by R.H. Popkin and Paul Dibon, the International Archives of the History of Ideas/Archives internationales d'histoire des idées, edited by Guido Giglioni and John Christian Laursen, with assistance of Former Director Sarah Hutton, publishes, edits and translates sources that have been either unknown hitherto, or unavailable, and publishes new research in intellectual history, and new approaches within the field. The range of recent volumes in the series includes studies on skepticism, astrobiology in the early modern period, as well as translations and editions of original texts, such as the *Treatise of the Hypochondriack and Hysterick Diseases* (1730) by Bernard Mandeville. All books to be published in this Series will be fully peer-reviewed before final acceptance.

Marta García-Alonso • John Christian Laursen
Editors

The Importance of Non-Christian Religions in the Philosophy of Pierre Bayle

Editors
Marta García-Alonso
Moral And Political Philosophy
UNED
Madrid, Spain

John Christian Laursen
Political Science
University of California-Riverside
Riverside, CA, USA

ISSN 0066-6610   ISSN 2215-0307 (electronic)
International Archives of the History of Ideas Archives internationales d'histoire des idées
ISBN 978-3-031-64864-9   ISBN 978-3-031-64865-6 (eBook)
https://doi.org/10.1007/978-3-031-64865-6

© The Editor(s) (if applicable) and The Author(s), under exclusive license to Springer Nature Switzerland AG 2024

This work is subject to copyright. All rights are solely and exclusively licensed by the Publisher, whether the whole or part of the material is concerned, specifically the rights of translation, reprinting, reuse of illustrations, recitation, broadcasting, reproduction on microfilms or in any other physical way, and transmission or information storage and retrieval, electronic adaptation, computer software, or by similar or dissimilar methodology now known or hereafter developed.
The use of general descriptive names, registered names, trademarks, service marks, etc. in this publication does not imply, even in the absence of a specific statement, that such names are exempt from the relevant protective laws and regulations and therefore free for general use.
The publisher, the authors and the editors are safe to assume that the advice and information in this book are believed to be true and accurate at the date of publication. Neither the publisher nor the authors or the editors give a warranty, expressed or implied, with respect to the material contained herein or for any errors or omissions that may have been made. The publisher remains neutral with regard to jurisdictional claims in published maps and institutional affiliations.

This Springer imprint is published by the registered company Springer Nature Switzerland AG
The registered company address is: Gewerbestrasse 11, 6330 Cham, Switzerland

If disposing of this product, please recycle the paper.

Contents

1 Introduction: Europe as a Religious Melting Pot.................. 1
  Marta García-Alonso

2 Beyond Conscience: Judaism in the Philosophy of Pierre Bayle 15
  Adam Sutcliffe

3 Philosophy and Its Islamic Moment: Pierre Bayle and Islam 41
  Pierre-Olivier Léchot

4 Bayle: Confucianism and China 71
  Marta García-Alonso

5 Bayle and Japan 99
  Fernando Bahr

6 Bayle's Reception of Greco-Roman Religion and Culture 125
  Parker Cotton

7 Bayle and the Ghosts of Mani and Zoroaster.................... 151
  Jean-Luc Solère

8 Bayle and the American and African Atheists 189
  John Christian Laursen

Index... 207

About the Editors

Marta García-Alonso is a historian of political ideas and has dedicated her research to religious reformers and French-speaking philosophers of the sixteenth and seventeenth centuries. She has written papers for *History of European Ideas, Intellectual History Review, History of Political Thought*, and coordinated works for Honoré Champion (Les Lumières radicales et la politique, 2017). Among her latest writings are "La hermenéutica bíblica hobbesiana del Leviatán," *Estudios Eclesiásticos* 98 (2023), pp. 305–337; "La Boétie, Étienne de," in M. Sellers and S. Kirste (eds.), *Encyclopedia of the Philosophy of Law and Social Philosophy*, Springer, Dordrecht, 2022; "Persian Theology and the Checkmate of Christian Theology: Bayle and the Problem of Evil," in W. Mannies, J.C. Laursen, and C. Masroory (eds.), *Visions of Persia in the Age of Enlightenment*, Oxford University Studies in the Enlightenment, Liverpool University Press (2021), pp. 75–100; "Calvin's Political Theology in Context," *Intellectual History Review* 31 (2021), pp. 541–556.

John Christian Laursen is Professor of the Graduate Division in the Department of Political Science at the University of California, Riverside. He has been writing on Bayle off and on since 1998, and his respect for Bayle has only grown over the years. His most recent edited books are *Clandestine Philosophy: New Studies on Subversive Manuscripts in Early Modern Europe, 1680–1823*, co-edited with Gianni Paganini and Margaret Jacob (University of Toronto Press, 2020) and *Persia in the Enlightenment*, Oxford University Studies in the Enlightenment, co-edited with Cyrus Masroori and Whitney Mannies (Liverpool University Press, 2021).

Chapter 1
Introduction: Europe as a Religious Melting Pot

Marta García-Alonso

Abstract Ethnographic knowledge is not merely used rhetorically or eruditely in the work of Pierre Bayle. Travel narratives describing non-Christian religions and cultures are the touchstone required for any claim that presents itself as genuinely universal in the seventeenth century. At the same time, they serve to demonstrate the particularity and relativity of many doctrines held to be universal. Many of Bayle's doctrines arise from the intellectual encounter between Christian doctrines and traditions foreign to Christianity. The knowledge of the other, the strange, the different is the *conditio sine qua non* that allows one to distance oneself from one's own traditions and values, a distance that is at the very origin of the philosophical thought that the Greeks already showcased.

Keywords Bayle ethnography · Transculturalism · Religion · Travel narratives · European traditions

This volume decenters the Christian roots and culture of early modern philosophy by bringing out the non-Christian roots and sources in the work of Pierre Bayle. Since Bayle was one of the most widely read and influential of the early modern philosophers, this should have the effect of decentering Christianity in our understanding of the history of early modern philosophy. Precisely because of the interest of Bayle and others in other religions, there has always been a wider set of religious and cultural influences in Europe than the assumption of exclusive Christian hegemony implies.

Thank you to John Christian Laursen for helpful suggestions on this introduction.

M. García-Alonso (✉)
Departamento de Filosofía Moral y Política, UNED, Madrid, Spain
e-mail: mgalonso@fsof.uned.es

© The Author(s), under exclusive license to Springer Nature Switzerland AG 2024
M. García-Alonso, J. C. Laursen (eds.), *The Importance of Non-Christian Religions in the Philosophy of Pierre Bayle*, International Archives of the History of Ideas Archives internationales d'histoire des idées 251, https://doi.org/10.1007/978-3-031-64865-6_1

This shift is important because there has long been an almost unanimous consensus in acknowledging the profound impact of Christianity on European culture and philosophy to the exclusion of almost everything else. From the widely embraced thesis of Carl Schmitt in *Political Theology* (1922/2007) suggesting that all political concepts of the modern state are secularized theological concepts, to the proposal by Jürgen Habermas of a post-Christian Europe (1994), there is a prevailing understanding that the Western tradition is not only deeply indebted to Christian traditions but that its concepts, doctrines, and authors form by far the most important part of a shared cultural heritage that all contemporary Europeans hold in common. For instance, a unique connection between freedom of thought, religious liberty, and Protestantism was championed by much of nineteenth-century German philosophy. It is worth noting that Fichte regarded the Protestant Reformation as the most potent force in German history, even considering the Reformers freethinkers. Notably, Hegel went further, equating the Reformation with modernity and as the engine of the Spirit's development. Even for Feuerbach, the Reformation was perceived as the catalyst for human liberation by converting theology into anthropology, thereby placing Christ in the position God held in medieval ontotheology. Even the skeptical Nietzsche, a critic of Christianity, while asserting that the Reformation hindered the German Enlightenment, found a place in his secular pantheon for Luther, crediting him as decisive in the development of the German language, of which the philosopher saw himself as a successor.

In a similar sense, Spain is often perceived as a nation whose essence is predominantly or exclusively Catholic. The memories of Jews and Muslims are primarily evoked to discuss their expulsion, in the face of the alleged necessity for the birth of the modern state, or to recall the margins of heterodoxy—a thesis championed by Marcelino Menéndez Pelayo in his *Historia de los heterodoxos españoles* (1880–1882). This very Spanish essence is paradoxically posited to explain the distinct Spanish Enlightenment or the country's delay in joining modernity—a recurring theme in twentieth-century philosophers like José Ortega y Gasset in his well-known analyses *La España invertebrada* (1921) and *La rebelión de las masas* (1929/2013). This is a modernity which the rest of Europe is believed to have embraced through Protestantism and its advocacy of freedom of conscience. In this view, modernity is fundamentally conceptualized as European, Protestant, white, and liberal.

These ideas were not limited to Europe. They travelled to the United States with Tocqueville. In *Democracy in America* Puritanism is presented as the true foundation of American democracy, positing the seventeenth-century religious conflicts as the genuine source of the idea of political freedom. Later on, Georg Jellinek argued in *General Theory of the State and Political Science* (1896) that the origins of human rights should be sought in Protestantism rather than the French Revolution.

The resonances of these genealogies have long reverberated among historians of ideas, leading to a genuine revival of the connection between Protestantism and political modernity. In this light, since George Williams published *The Radical Reformation* (1962), focusing on the political influence of the Anabaptist tradition, there has been a growing interest in figures not directly related to the mainstream

Calvinists and Lutherans of the Reformation. John Pocock emphasized the role of Arminians in the Enlightenment of the Netherlands (1999); Jonathan Sheehan revisited the importance of the relationship between biblical hermeneutics and the Enlightenment (2005); Michael Gillespie argued that the Enlightenment should be interpreted based on the ontological sources of the nominalist tradition, in which the Reformation is inscribed (2008); and Teresa Bejan's *Mere Civility: Disagreement and the Limits of Toleration* (2017) focuses exclusively on Protestant Englishmen (Roger Williams, Hobbes, and Locke) in order to discuss the rise of European toleration. All agree, in one way or another, on pinpointing Christian sources at the origins of modernity and the Enlightenment.

Almost everywhere in the study of Euro-American history the focus is on Christianity, not the other world religions or atheism, which invariably labels those religions as undesirable and enemies of order. Despite the recurring presence of atheists and atheism in a wide variety of theological, philosophical, and literary texts, it is only recently that they are being brought back to consideration as important elements of our common history (Israel 2001; Paganini et al. 2020).

There are larger implications here. If these Christian genealogies of our most important political doctrines, ethical values, and epistemic philosophies were the only genealogies we have, it would be unjustifiable to reject inclusion of Christianity in the European Constitution. This proposal, influenced by John Paul II, was put forward by certain conservative groups, although it faced significant opposition (Zucca 2012). In this proposal, Judaism and Islam were mere traveling companions, not always welcomed or understood. Confucianism and Buddhism are even more distant from our European history. In contrast with that, in this volume, we aim to demonstrate that the relationship between European philosophers and other cultures and religions is complex enough to reconsider that privileged link between Christianity and the European intellectual tradition that is often so uncritically accepted. Travel books written by diplomats, missionaries, or merchants unveiled a new world of doctrines, practices, and stories that fertilized the emergence of new ideas in the West and allowed for a critical reassessment of the well-known Greco-Latin classical traditions (Withers 2007). America, Africa, and Asia were depicted as more than just territories for political colonization, serving as sources of new commodities, including human resources, and areas for mining and cultivating new crops for Europe. They also represented fertile grounds for the emergence of new ideas and political institutions.

As we will see in this volume, from the seventeenth century onwards, knowledge, traditions, and cultural values from Asia and America were used to challenge the old Christian doctrines that had hitherto been taken for granted. This encounter, as argued by Paul Hazard (1934), destabilized old Europe in more than one way. However, this cultural clash should not be understood merely historically, as an event that occurred at a specific time affecting only a specific part of our tradition. We believe it should also be perceived as a broad epistemological paradigm shift, affecting the future and the new lines of thought it uncovered, which are still being explored. The anthropological data obtained from the knowledge of other lands eventually became part of the ethical and political discourse of our philosophers, historians, and literati, similar to the way that empirical data from those continents were incorporated into medicine, botany, or navigation. In some sense, it could be argued that ethnographic reports from other cultures and religions occupied in philosophical discourse from the seventeenth century onwards the place held by revealed data in theology. In many ways, they were the material with which ethical and political thought was built; the source from which new ideas sprang. For this reason, one might say that knowledge of other cultures and religions was the critical mass needed to reflect on the particularities of Christian Europe itself. This was clearly the case with Pierre Bayle.

Perhaps the most comprehensive and detailed recent examination of Bayle's context and important aspects of his philosophy is Gianni Paganini's *De Bayle à Hume* (2023). It does not, however, discuss Bayle's understanding of and evaluation of non-Christian religions in any systematic way. For example, Paganini treats Bayle's view of the Zoroastrians in Bayle's *Conversations between Maxim and Themiste* in two pages, whereas Jean-Luc Solére's chapter here is much more thorough. Similarly, there is not much specific attention to China, Japan, America, or Africa and some other places in Paganini's work. A reading of the present book would help broaden the analysis to include Bayle's perspectives on many non-Christian religions, and how they affect his views on Christianity and atheism.

In this volume, we attempt to show how ethnographic knowledge is not merely used rhetorically or eruditely in the work of Pierre Bayle, but functions as a testing ground for any doctrine, idea, or value that is presumed or proposed to be transcultural. We will see how the philosopher from Rotterdam used Greco-Roman culture to compare ancient virtue with the vices of his contemporary co-religionists and employed his vast knowledge of the ancient pantheon as a means to criticize certain aspects of Christian dogma. Likewise, he spoke of the Muslim tradition as a culture that treasured and brought Greek philosophy to the West, without which freedom of thought in Europe would not have flourished, deeply linking Muhammad's religion to European culture. In the same way, he extracted examples from distant lands in order to validate doctrines that were thought to be exceptional until then. For example, Bayle associated Buddhism with Catholic quietism and Spinozism. From then on, the defence of a possible human perfectibility became a belief common to East and West, using Japan as a bridge. It is true that Africa and America receive very limited attention in his work, but their mention was essential to prove that there are societies that can prosper, or at least coexist, without any religious belief. The Christian bond between religion and politics faded in the face of these anthropological examples. Similarly, Confucianism was seen by Bayle as a civic philosophy that underpinned a government of atheists, essential to explain the flourishing of Chinese civilization. Equally interesting is the lesson that can be drawn from his knowledge of the Jewish tradition, as his study of it showed the limits of the doctrine of freedom of conscience. Bayle did not advocate a universal extension of his doctrine of religious toleration but confined its use and validity to the acceptance of the Christian framework that gave it meaning.

Therefore, as the texts compiled in this volume show, it can be argued that many of Bayle's doctrines arise not so much –or not only– as an internal development of a heterodox Christianity (Labrousse 1964; Bost 2006), but as a result of the intellectual encounter between Christian doctrines and traditions foreign to Christianity. This encounter does not imply a major civilizational clash. The knowledge of the other, the strange, the different is the *conditio sine qua non* that allows one to distance oneself from one's own traditions and values, a distance that is at the very origin of the philosophical thought that the Greeks already showcased. And while it is true that travel narratives describing non-Christian religions and cultures are the touchstone required for any claim that presents itself as genuinely universal, at the same time they serve to demonstrate the particularity and relativity of many of the doctrines held to be universal up to that point.

We will see that Bayle's approach to other cultures and religions is not entirely faithful to the religious and cultural traditions he surveys, nor is it always free from prejudice. On the one hand, his knowledge of Buddhism, Confucianism, Persian religions, or American atheism is mediated by the authors of the travel books and their own commercial, evangelical, or political interests. On the other hand, Bayle himself brings up any custom or religious ritual based on his ideological interests, his ethical values, or simply out of sheer intellectual opportunism, as the chosen formula to expose an adversary in a specific debate. Bayle does not reflect on objective data for its own sake but interprets and rereads documents and facts in light of his philosophical concerns.

As far as Judaism is concerned, Richard Popkin's scholarly endeavors opened avenues for considering Bayle as someone who, beyond academic curiosity, might have identified with Jewish philosophy to a significant extent. He explored the nuanced interplay between Bayle's philosophical writings and Jewish thought. He found that Bayle may have felt an inherent connection to Jewish beliefs, a notion that highlights a deep, though perhaps symbolic or ideological, association with Jewish teachings (Popkin 1996). In contrast, Miriam Yardeni argued that although Bayle did not share the prevalent deep-seated animosity towards Jews, well-known among French Calvinist pastors and theologians, in some respects he was anti-Jewish, viewing Judaism as fundamentally lacking in moral substance (Yardeni 1980).

With more nuance, in his chapter Adam Sutcliffe demonstrates that the portrayal of Judaism in Bayle's work was intricate, as evidenced by his studies on biblical exegesis and his profound interest in specific Jewish individuals such as Spinoza. Spinoza's combination of atheism with a geometrical approach to ethics struck Bayle as not just nonsensical, but also as deeply overconfident and rigid. The aspect of Spinoza's philosophy that Bayle found most troubling and perilous was its rationalistic dogmatism. Above all, Sutcliffe highlights a crucial aspect that has been overlooked by scholars of the philosopher from Rotterdam: Judaism reveals the limits of Bayle's notion of freedom of conscience. The paradox is that Bayle staunchly upholds the rights to individual conscience as absolute and unquestionable, yet at the same time refrains from extending this to the Jews. Jews who abandon Judaism in pursuit of their own conscience are deemed bad Jews. It is as straightforward as that.

In Bayle's perspective, Jews should epitomize unwavering adherence to the Mosaic Law. The idea that Judaism as a faith leaves no room for individual conscience was a cornerstone belief, not just for Bayle but also for many in his broader intellectual environment. He contended that every individual is ethically obligated to heed the voice of their conscience, irrespective of its origins or its religious underpinnings. Yet, given that Jews are direct beneficiaries of God's explicitly revealed and binding Word, their spiritual and intellectual existence was distinctively different. There was no tug-of-war between reason and faith for Jews, or at least there should not be since every facet of Jewish tradition and law was directly under the dominion of divine command. In their situation, the prerogatives of personal conscience were overshadowed by the supremacy of the Mosaic proclamation, which Bayle underscored as 'a universal edict for the Jews'. Therefore, a Jew following his conscience is effectively a wayward Jew. If he wishes to adhere to the law of conscience, he ought to embrace Christianity, Bayle thought.

The presence of Islam and its impact on Bayle's work has not been extensively examined by specialists. Exceptionally, Mara van der Lugt (2017) has recently revisited some of Bayle's texts concerning the topic of sexuality and war. In van der Lugt's view, Bayle addressed the sexual ethics in Islam, questioning the narrative that Mahomet's personal desires shaped Islamic precepts and discussing the Prophet's relations with his wives and his promises of a sensual paradise. Furthermore, Bayle critiqued the position of women under Islamic law, highlighting its disadvantages and arguing for a proto-feminist perspective that recognized women's rights and condemned their mistreatment. He called for a practice of religious tolerance and a critical re-examination of the roles of war and sex in the histories of both Islam and Christianity (van der Lugt 2017).

Despite its significant interest, van der Lugt's work only addresses some very specific aspects of Islam. In the study presented here by Pierre-Olivier Léchot, however, the analysis of the presence of Mohammed's religion in Bayle's work is much more comprehensive and detailed. Léchot argues that even though Bayle was not directly an Arabist, he certainly paid keen attention to the academic works related to Islam and the broader Orient. Yet, the portrayal of the Prophet that one finds in Bayle's writings is predominantly negative. Bayle perceived Muhammad as a deceiver, labelling him a "charlatan" and a "counterfeit prophet". This characterization of Muhammad as a mere charlatan was not the only perspective available. Casting the Prophet as a charlatan meant sidelining other interpretations of him as a menacing zealot, a tool of evil, or even an embodiment of the Antichrist, even though these views were heavily emphasized in medieval traditions. It is important to note that Bayle characterized Islam as simultaneously aggressive (in its outward spread) and accepting (in handling religious diversity within its dominions). Such a portrayal enabled him to emphasize the doubly contentious nature of Catholic practices: the latter are not only as reprehensible as the methods of the 'Turks' (Bayle alludes to the crusades), but surpasses them, given that Muslim dominions permit diverse religions to coexist and the Catholics do not. More critically, Bayle believed that a lineage of free-thinking persisted within Islam, a lineage that culminated in figures like Spinoza and can be traced back not just to Averroes but also to other lesser-known personalities in the Arab-Muslim tradition. Thus, for Bayle it was both the expansionist strategies and the relative tolerance of Muslims that ensured the continuity of certain Greek philosophical schools that championed views deemed heretical by Christians. The latter, in contrast, had often suppressed or even eradicated them.

Bayle's attitude China has consistently captivated scholars' interest, particularly in the context of a society perceived as atheistic, a theme evident in the works of recent scholars Gianluca Mori and Jean-Michel Gros. For instance, Mori speculates that Bayle's writings do not necessarily prioritize the historical presence of atheism but rather construct a philosophical allegory surrounding the concept of sophisticated atheism. This metaphor elevates atheism to a high intellectual plane, portraying it as the culmination of a logical deconstruction of religious belief. However, this atheistic perspective is described by Bayle as being beyond the grasp of the masses. Mori explains that Bayle does not embody the role of a progressive philosopher with an agenda to usher the masses towards a secular awakening. Rather, in Mori's interpretation Bayle appears to suggest that religion, deeply rooted in human emotion, is an indelible aspect of human culture, much like a persistent disorder. According to Mori, for Bayle the ubiquity and persistence of religious sentiment resemble an endemic malady that cannot be eliminated but can be managed through tolerance (Mori 2011). In a similar sense, Jean-Michel Gros argues that Bayle thought that civil peace was only possible if the state were functionally atheistic. Bayle believed that the state should not interfere in matters of faith and should not take sides in confessional conflicts, but rather should maintain civil peace amidst religious disorder. In that case, a society of atheists would be outwardly indistinguishable from a society of Christians because religion does not contribute to the improvement of the morals of its adherents and, therefore, does not strengthen the social bond (Gros 2004).

Other scholars, such as Gianni Paganini (2009), have found in Bayle's texts related to China a way to engage with the complex concept of atheism itself. Paganini discusses Bayle's nuanced understanding of atheism and outlines how contemporary research has shifted from a strict metaphysical notion of atheism — defined as the denial of the existence of a first cause or supernatural agent— to a broader understanding that encompasses a range of non-conformist religious phenomena, heterodoxy, or impiety. Bayle's contribution to this discourse is significant; he acknowledges a spectrum of atheistic belief, distinguishing between dogmatic or affirmative atheism and skeptical atheism. Paganini argues that Bayle's concept of atheism is both rigorous and expansive, offering a modern notion of skeptical atheism. This concept is not just an idle classification but has implications for understanding Bayle's own perspectives on atheism and the larger historical context of religious belief and skepticism. Paganini suggests that Bayle's atheism is not a simple negation of theism but a more complex position that invites deeper exploration into the motives and content of atheistic thought as expressed by Bayle and subsequent thinkers (2009).

However, no one has ventured into the study of the influence of China in a broader sense, nor do they address Bayle's own concept of Confucianism, with the exception of Simon Kow, with whom Marta García-Alonso engages in debate in his volume. Kow argues that Montesquieu's account of Confucianism in *The Spirit of the Laws* was partially influenced by his critique of Bayle's position on the role of religion in society. While Bayle posited that a society of atheists could be morally decent, Montesquieu challenged this separation of religion from morality and politics, seeing Confucianism as both a religion and a moral and political code, integral to Chinese society (Kow 2017).

In her chapter García-Alonso offers an analysis of the importance that China holds for Bayle, not only in relation to his concept of atheism or the concept of natural religion, but in his understanding of Confucianism as a civil religion. She observes that in Bayle's opinion China had a high level of complexity and development, in evident contrast to America and Africa. Thus, it is difficult to support the idea that morality is tied to religiosity, as theologians do, when we have knowledge of a people who have lived for centuries without knowing Christianity and yet they have impressive cultural and moral achievements. Besides, the atheism of the Chinese is not merely a negative absence of belief, like that of the Americans or Africans, but rather a conscious rejection of all *ontotheology*. China has philosophical atheism (positive atheism), not just cultural atheism (negative). And that atheism originated from the teachings of Confucius. In China, Confucius is not worshipped as a divine figure but rather respected as a philosopher known for his ethical teachings. Contrary to the Jesuit perspective in Bayle's time, there was no inherent religion; the presence or absence of religious beliefs in certain societies could be attributed to education and tradition, Bayle thought. Furthermore, China, with its Confucian governance devoid of theism, offered Bayle a case to emphasize that the foundation of social cohesion and formation of political communities do not necessarily rely on religion. What is essential is the civil law, which should refrain from intruding upon religious convictions.

As highlighted by García-Alonso, Bayle understood from the Jesuits that the Chinese followed both Buddhism and Taoism; hence, he did not conflate atheistic governance with a purely secular state. Nonetheless, Bayle might have perceived China as a prototype of a state without an official religion. While it mandated Confucian civic events, he saw these more as public ceremonies rather than religious rites. China is the quintessential example of how a country can be governed without relying on religious legitimization. In that sense, García-Alonso argues that Bayle's arguments are not only significant for their historical and philosophical implications but also have contemporary relevance. Today, there are ongoing debates about the role of religion in politics and the degree to which it should influence public policy. Bayle's argument provides a crucial perspective that can help inform these debates, highlighting the possibility of rejecting any religious legitimization of politics in a religiously plural society.

In contrast, neither Japan nor Buddhism in Bayle's work are subjects that have received much attention from specialists. Certainly, Japan is often cited as an example alongside China in the context of Bayle's discussions of political tolerance and sedition. It is known that Bayle cautions the Japanese against the Christians and their political doctrines. In his chapter on Japanese religions, Fernando Bahr illustrates that Bayle's primary concern was not directly documenting Buddhism, but rather highlighting parallels between certain tenets held by distant bonzes and the ideas rooted in Spinoza's philosophy or the principles of Catholic "quietist" mystics—both of which were pivotal intellectual discussions in the Republic of Letters during his era. Bahr shows that Bayle saw striking similarities between the misguided objectives of Christian mystics and Eastern philosophers in their attempts to bridge the gap with the divine.

However, from a Calvinist perspective in Bayle's time, no virtuous deed could ever lessen the vast chasm separating God from His creations, whether achieved via prayer—as posited by Catholic quietists—or through meditation—as advocated by Buddhists. Equally significant is the conclusion drawn from such misguided beliefs: any theory that undermines God's sovereignty aligns with atheism in propounding one of the gravest and most definitive errors imaginable—the prospect of a human soul untouched by sin. Put simply, both quietists and Buddhists uphold the notion of human flawlessness. This mirrors Spinoza's stance in his *Ethics*, where he posited that beatitude lies in recognizing or understanding our innate perfection. Thus, in Bayle's eyes, quietism (whether of the East or West) bore connections to Spinozism, which he opposed. While Buddhists were unfamiliar with the concept of original sin and even denied the existence of evil, quietists believed that those deemed "perfect" could achieve unity with God by allowing the divine to reside within them. Spinoza proposed that intuition was better than reason, culminating in *amor Dei intellectualis*. Bayle strongly opposed both of these perspectives.

In turn, Greco-Roman religion is usually associated with the discourse on paganism, which has indeed been addressed by nearly all scholars studying Bayle. However, classical culture in Bayle has only been superficially explored, apart from Elisabeth Labrousse. She highlighted that while many humanists valued ancient cultures primarily for their aesthetic appeal, Bayle did not share this sentiment. He expressed concern over the declining interest in Greek studies of the French, a trend not seen in England. However, Bayle's advocacy of Hellenism was not for its potential to delight, but as a field of scientific study where the French previously excelled. Bayle's focus was on the philosophers rather than the poets, prioritizing conceptual precision in translations over poetic elements like rhythm or musicality. Bayle sought insights from antiquity through the lens of historians and philosophers, rather than poets. Furthermore, Labrousse established that Bayle criticized the ancient historians for their lack of precision in timelines, while recognizing their keen insights into human motivations. Certainly, the historians lacked precision in chronological matters, but on the other hand, they expertly probed human motives and provided purely natural explanations for events. As for the philosophers, they explored the coherence of the world (Labrousse 1996).

In this volume, Parker Cotton continues the work of Labrousse and demonstrates that for Bayle, Greek and Roman Antiquity was a continual point of reference. While Rome stood as a paradigm for governance, military might, and eloquent writing, it was the Hellenic realm that inspired innovative philosophical thinking and poetic reverie. In Bayle's narratives, there is a blending of these two ancient worlds, especially when recounting their intertwined religious pantheons, though distinctions become pronounced when he mentions Rome's god-like emperors. In the context of the seventeenth century, intellectuals of the Republic of Letters often likened themselves to Roman poets and orators, subtly acknowledging the sophisticated conduct epitomized by the ancient Romans. This was a world where intellectual freedom reigned supreme, allowing each thinker to challenge the others with truth and logic in the pursuit of knowledge. For those well-versed in the discourse of ancient Rome, this was the ideal to strive for. In contrast to the laudable portrayal of Roman figures, Bayle criticized the heathen deities. He found a disconnect between morality and the teachings of the pagan faiths. While religious tenets advocated either virtue or vice, mankind's moral compass remained fairly consistent. According to Cotton, the entries on Greco-Roman topics in Bayle's works enabled him to probe modern religious morals, yet with a comfortable buffer. However, the comparative utility of these entries is multifaceted, contingent upon Bayle's immediate objective or thematic emphasis. They enrich the complex yet enthralling religious discussions in the *Dictionary*, defying simplistic conclusions and rigid interpretations.

Unlike what happens with classical culture, Bayle's view of Persia has always been of interest to scholars, albeit indirectly, as it paves the way for discussion of the problem of evil. In recent years, Michael Hickson has devoted important scholarship to this topic, bringing out Bayle's profound skepticism concerning theodicy (Hickson 2011). Central to Bayle's critique is the *reductio ad malum* method that reveals the internal inconsistencies within monotheistic attempts to vindicate a benevolent God's causal involvement with evil, both physical and moral. Hickson further contends that Bayle's skeptical stance should be understood not as indicative of atheism but as part of a broader Protestant intellectual milieu that tolerated, if not embraced, such critical inquiry. Moreover, in his work, he points out that Bayle's treatment of the persecutor paradox reveals a fundamental tension in Bayle's thought: the moral imperative to follow one's conscience leads to a philosophical impasse when applied to persecutors, who, by the same logic, would be morally justified in following their conscience toward persecution. Bayle's eventual turn to a skeptical defense of toleration indicates his recognition of this tension and represents a philosophical shift that anticipates his later fideistic tendencies.

In contrast, in a recent chapter Marta García-Alonso explored Bayle's treatment of Persian thought in relation to Christian theology (García-Alonso 2021). She suggested that Bayle's engagement with the problem of evil was not just an academic exercise, but a philosophical critique aimed at undermining the necessity of theology itself. Bayle's work called into question the use of philosophy in religious matters and critiqued the Calvinist and Catholic theologies of his era, particularly targeting the theological constructs represented by Augustine of Hippo. García-Alonso argued that Bayle's treatment of the problem of evil served as a polemic against the application of philosophy to religious affairs and that Bayle's objective was to question the misuse of philosophy rather than to cast doubt on the possibilities of philosophy itself. Bayle's work is seen as a critique of the illegitimate use of philosophy in religious contexts and to direct criticism not just at the Calvinist theology of his time but at the essence of Christian theology as a whole. She interprets Bayle's extensive treatment of the problem of evil as an indirect attack on the need for theology, positioning Bayle as a critical voice that advocated separating religious belief from philosophical rationalization, thus favoring a personal and subjective approach to faith over objective theological mandates (García-Alonso 2021).

In his chapter, Jean-Luc Solère delves into these debates by incorporating a systematic study of the dialectical context in which the discussion about evil unfolds. He points out that comprehending the historical backdrop of religious disputes is crucial for deciphering Bayle's intricate strategy concerning theodicy. Solère illustrates that as early as the sixteenth century Catholics were quick to denounce Lutheranism and Calvinism, accusing them of reviving Manichaeism. This accusation stemmed from the belief that Lutheranism and Calvinism denied human freedom in the present life and asserted that, without God's grace, humans could only sin, similar to being under the dominion of the Evil Principle. This accusation of Manichaeism was not limited to Catholic criticism but also played a prominent role in Protestant controversies. Pierre Jurieu, a staunch Calvinist, leveled the accusation of Manichaeism against Luther, while Lutherans directed it at the Calvinists.

Likewise, in response to the recurring accusation of Manichaeism, Solère argues that Bayle employed a strategy known as *rétorsion*, involving a counter-accusation or redirecting of arguments against the accusers. He demonstrated that they too could be subject to the same objection they had raised. This tactic allowed the defenders of each position to remain steadfast since the opponents' positions suffered from the same flaws. Consequently, in debates between Christians and *diarchists*, the only recourse was to retreat to the safe haven of Revelation while acknowledging that reason couldn't resolve the Manicheans' objections. However, it is essential to note, according to Solère, that Bayle's stance was not merely blind fideism. It was grounded, in part, on the rational notion that God always acts rightly, and faith entered the equation to provide a content to which this principle could be applied. Consequently, it was in everyone's best interest to avoid delving into the issue of God's responsibility regarding evil and instead let it remain unresolved. This approach aimed to prevent fruitless controversies that had plagued Christianity, and to promote tolerance for all people of goodwill.

Last but not least, Bayle exhibits a comparatively lesser interest in America and Africa, with more attention to the Americas than Africa. His portrayal of Africa is inextricably linked to the historical narratives of European exploration and the ensuing commercial pursuits. Bayle's examination of the Americas extends beyond mere geographical or cultural descriptions, delving into the realms of religion and colonialism. He contends that atheistic societies, contrary to prevailing European beliefs, could exhibit moral and virtuous characteristics, potentially surpassing those of idolaters or even Christians. This assertion, grounded in traveller accounts from Canada and other parts of the Americas, and Africa, serves not only as a critique of European moral superiority but also as a strategic instrument in the broader discourse on atheism and ethics.

Bayle's thoughts about atheism in American and African cultures have been studied by scholars Juliette Charnley (1990) and Corinne Bayerl (2007). Bayerl pointed out that Bayle suggested that both some forms of atheism in indigenous societies and the modern critical mind were less likely to make some kinds of errors than religious people were. She calls this "immunity from error", which may be going too far, especially for a thinker like Bayle who found error almost everywhere, but there can be no doubt that Bayle is challenging traditional assumptions about the relationship between religion and civilization. She maintains that Bayle's argument is not a straightforward endorsement of atheism but rather a nuanced exploration of the role of belief and unbelief in historical and moral development. He did not see religious belief or atheism as the original state of mankind but instead viewed all societies as already corrupted. His approach was also marked by a refusal to idealize early stages of human societies in comparison to later forms. Moreover, as Charnley pointed out, Bayle's commentary on the Americas includes a critical perspective on the forced conversion of indigenous populations by European colonizers. He underscores the superficial nature of these conversions, lacking in genuine persuasion or enlightenment, thus reflecting a critique of European colonial practices. His remarks on the perceived absence of 'civilization' in the Americas further exemplify the era's Eurocentric viewpoints (Charnley 1990).

In his chapter in this volume, John Christian Laursen delves further into Bayle's writings, especially in the *Continuation of the Diverse Thoughts on the Comet* (1704), where Bayle provides ethnographic evidence of atheism among various cultures, ranging from the Americans and Africans to the Chinese and the inhabitants of the Mariana Islands in Oceania. Notably, Bayle did not limit his observations to highly advanced societies, indicating that atheism can be present even in less developed contexts. Bayle critically evaluated common misconceptions about so-called "barbarian" societies. He acknowledged that there seemed to be differences in terms of technological advances, aggressive tendencies, and moral values among these populations. However, he firmly contested the prevailing assumption that recognizing and venerating physical objects like rocks, woods, or streams as divine is an automatic indication of lower cognitive faculties than those who venerate a more abstract god. As Laursen points out, it is possible that Bayle's lesser interest in the Americans and Africans may be attributed to the lack of a written culture that he could read, either on his own or in translation.

It is evident that the knowledge and interest that Bayle demonstrated toward non-Christian religions and cultures varied widely. Yet, his work in these areas was not just an exercise in amassing erudite information to embellish his writings. His works and ideas were shaped and transformed in light of the ethnographic data he continuously gathered. In a similar vein, his philosophical endeavors should be recognized as pivotal in outlining the pathway any philosophy should undertake if it seeks to liberate itself from the overpowering grasp of theology. In his pursuits, Bayle ventured beyond mere sacred scriptures and the prevailing theological interpretations of Greco-Latin culture. His aim was to present an intellectual alternative to his contemporaries. Through his lens, we are exposed to nuanced interpretations of the Greco-Latin and Jewish Mediterranean traditions. He beckoned us to delve deeper into Eastern realms, exploring Islam, Japanese Buddhism, Chinese Confucianism, and Persian Zoroastrianism. And he extended a compassionate, non-condemnatory viewpoint towards the atheist societies he identified in regions of the Americas and Africa.

As this volume aims to demonstrate, one of Bayle's overarching goals was to advocate an enhanced understanding and appreciation of foreign cultures in order to understand our own. This served as a fountainhead of novel insights, equipping his readers with a reinforced skepticism, especially towards the claims of absolute truth made by Christian theology. Such skepticism, it is worth noting, is perhaps the only mechanism that effectively counters the alleged monolithic essences that religions assert as their own. It challenges the entrenched belief that intellectual, religious, or political traditions are bound to honor an inherent purity and uniformity. The ability to distance oneself from ingrained beliefs, amplified by the acquisition of knowledge regarding unfamiliar customs, is the very essence of critical thinking. Such thinking resolutely militates against both philosophical dogmatism and religious orthodoxy. Bayle stands tall as one of the most illustrious advocates of critical thinking. This perspective persists as the sole remedy, even in contemporary times, against the pitfalls of philosophical and religious parochialism, and serves as the bedrock upon which we can foster communal coexistence.

Bibliography

Bayerl, Corinne. 2007. Primitive atheism and the immunity to error: Pierre Bayle's remarks on indigenous cultures. In *Religion, ethics, and history in the French long seventeenth century. La religion, la morale, et l'histoire à l'âge classique*, ed. William Brooks and Rainer Zaiser, 15–28. Oxford/Berlin: Peter Lang.
Bejan, Teresa. 2017. *Mere civility: Disagreement and the limits of toleration*. Cambridge: Harvard University Press.
Bost, Hubert. 2006. *Pierre Bayle*. Paris: Fayard.
Charnley, Juliette Joy. 1990. *The influence of travel literature on the works of Pierre Bayle with particular reference to the Dictionnaire historique et critique*. Durham theses. Durham University. http://etheses.dur.ac.uk/6574/
García-Alonso, Marta. 2021. Persian theology and the checkmate of Christian theology: Bayle and the problem of evil. In *Persia and the Enlightenment*, ed. C. Masroori, W. Mannies, and J.C. Laursen, 75–100. Oxford University studies in the Enlightenment. Liverpool: Liverpool University Press.
Gillespie, Michael. 2008. *The theological origins of modernity*. Chicago: University of Chicago Press.
Gros, Jean-Michel. 2004. Bayle et Rousseau: société d'athées et/ou religion civile. In *Pluralismo e religione civile. Una prospettiva storica e filosofica*, ed. G. Paganini and E. Tortarolo, 124–138. Milano: Mondadori.
Habermas, Jürgen. 1994. *Postmetaphysical thinking*. Cambridge: MIT Press.
Hazard, Paul. 1934. *La Crise de la conscience européenne, 1680–1715*. Paris: Boivin.
Hickson, Michael. 2011. Reductio ad Malum: Bayle's early skepticism about theodicy. *Modern Schoolman* 88 (3/4): 201–221.
Israel, Jonathan. 2001. *Radical Enlightenment: Philosophy and the making of modernity, 1650–1750*. Princeton: Princeton University Press.
Kow, Simon. 2017. *China in early Enlightenment political thought*. New York: Routledge.
Labrousse, Elisabeth. 1964. *Pierre Bayle: Hétérodoxie et rigorisme*. Vol. II. Netherlands: Springer.
———. 1996. Bayle et les deux Antiquités. In *Conscience et conviction. Études sur le XVIIe. Siècle*, 139–147. Paris/Oxford: Universitas-Voltaire Foundation.

Menendez Pelayo, Marcelino. 1880–1882/2003. *Historia de los heterodoxos españoles.* Alicante: Biblioteca Virtual Miguel de Cervantes. https://www.cervantesvirtual.com/obra/historia-de-los-heterodoxos-espanoles/

Mori, Gianluca. 2011. Religione e politica in Pierre Bayle: la 'società di atei' tra mito e realtà. In *I filosofi et la società senza religione*, ed. M. Geuna and G. Gori, 41–60. Boulogna: Società il Mulino.

Ortega y Gasset, J. 2013. *La rebelión de las masas*, ed. D. Hernández. Madrid: Tecnos.

Paganini, Gianni. 2009. Pierre Bayle et le statut de l'athéisme sceptique. *Kriterion* 120: 391–406.

———. 2023. *De Bayle à Hume: Tolérance, hypothèses, systèmes.* Paris: Champion.

Paganini, G., Margaret Jacob, and J.C. Laursen, eds. 2020. *Clandestine philosophy: New studies on subversive manuscripts in early modern Europe, 1620–1823.* Toronto: University of Toronto Press.

Pocock, John G.A. 1999. *Barbarism and religion, vol. 1, The Enlightenments of Edward Gibbon.* Cambridge: Cambridge University Press.

Popkin, Richard H. 1996. Pierre Bayle and the conversion of the Jews. In *De l'Humanisme aux Lumières: Bayle et le protestantisme*, ed. Michelle Magdelaine et al., 635–643. Oxford: Voltaire Foundation.

Schmitt, Carl. 2007. *Political theology.* G. Schwab (translator), B. Strong (foreword) and L. Strauss (notes). Chicago/London: The University of Chicago Press.

Sheehan, Jonathan. 2005. *The Enlightenment Bible: Translation, scholarship, culture.* Princeton: Princeton University Press.

Strauss, Leo. 2013. *On Maimonides: The compete writings.* Chicago: University of Chicago Press.

Tocqueville, Alexis. 2010. In *De la démocratie en Amérique*, ed. I.P. Raynaud. Paris: Flammarion.

van der Lugt, Mara. 2017. The body of Mahomet: Pierre Bayle on war, sex, and Islam. *Journal of the History of Ideas* 78: 27–50.

Williams, George. 1962. *The radical Reformation.* Philadelphia: Westminster Press.

Withers, Charles W.J. 2007. *Placing the Enlightenment: Thinking geographically about the age of reason.* Chicago: University of Chicago Press.

Yardeni, Miriam. 1980. La vision des juifs et du judaïsme dans l'oeuvre de Pierre Bayle. In *Les juifs dans l'histoire de France*, ed. M. Yardeni, 86–95. Leiden: Brill.

Zucca, Lorenzo. 2012. *A secular Europe: Law and religion in the European constitutional landscape.* Oxford: Oxford University Press.

Chapter 2
Beyond Conscience: Judaism in the Philosophy of Pierre Bayle

Adam Sutcliffe

Abstract Interpretations of Pierre Bayle's attitude to Judaism differ enormously, mirroring the more broadly divided nature of recent Bayle scholarship. It is meaningless, this essay argues, to categorise Bayle as either an 'antisemite' or a 'philosemite'. Jews occupy a uniquely anomalous and complicated place in his work because they are figure centrally in his explorations of the philosophical question that most preoccupied him: the inescapable incommensurability of faith and reason. The ensnarement of Jews and Judaism is here investigated across four key areas: Bayle's approach to the Old Testament (or Jewish Bible); his coverage of early modern Jews in his *Dictionnaire*; the place of Judaism in his arguments for toleration and the rights of conscience; and the persistence of Judaic themes in his most intensely contested late writings.

Keywords Pierre Bayle · Judaism · Conscience · Toleration · Spinoza · Jean Le Clerc

Scholarly assessments of Pierre Bayle's attitude to Judaism reflect in microcosm the divided state of Bayle interpretation in general. Just as critics cannot agree on whether Bayle was fundamentally a secular rationalist, a conflicted but committed Protestant, or the holder of some intermediate or hybrid position between these two extremes, a wide range of characterizations have also been put forward of the underlying nature of Bayle's approach to Judaism, Jewish history and Jews. According to Richard Popkin, a pronounced affinity with Judaism is apparent in Bayle's writings. In his inner heart, Popkin speculated in the introduction to his selective translation of the *Dictionnaire*, Bayle might have been 'a Judaizing Christian, or a genuine Judaeo-Christian, or even a secret Jew' (1991, xxvi). Miriam Yardeni, in contrast,

---

A. Sutcliffe (✉)
Department of History, King's College London, London, UK
e-mail: adam.sutcliffe@kcl.ac.uk

© The Author(s), under exclusive license to Springer Nature Switzerland AG 2024
M. García-Alonso, J. C. Laursen (eds.), *The Importance of Non-Christian Religions in the Philosophy of Pierre Bayle*, International Archives of the History of Ideas Archives internationales d'histoire des idées 251, https://doi.org/10.1007/978-3-031-64865-6_2

has argued that Bayle was in essence 'anti-Jewish', because he regarded Judaism as inherently morally deficient. While not finding Bayle personally guilty of antisemitism, Yardeni nonetheless charged him with opening up an intellectual road towards this prejudice, along which, on her account, Voltaire and d'Holbach later travelled without hesitation (1980). Lorenzo Bianchi, in a more measured essay on the subject, has argued that Bayle's critiques of Judaism, because they are historical rather than racial, are not 'radically anti-Jewish', and are in any case outweighed in significance by his commitment to universal toleration (1994).

How should we make sense of these widely diverging assessments? The place of Judaism in Bayle's thought and work has been an area of controversy because the stakes on the topic are high, and the broad terrain is confused. The eighteenth-century Enlightenment is widely regarded as culminating with the extension of political rights to religious minorities in the wake of the French Revolution, when the case of the Jews, both immediately in 1789 and over the course of the extremely protracted ensuing process of what became known in the 1830s as 'Jewish Emancipation', gave rise to uniquely intense discussion and debate. Jewish political inclusion was closely followed, though, by the rise of modern political antisemitism. In the light of this, the Enlightenment has been interpreted as the indispensable origin of the ideas that enabled Jewish inclusion and cultural flowering in the modern era, but also as an intellectual bridge between the religious antisemitism of pre-modern Europe and the later rearticulation of this prejudice in secular form.[1] As a foundational thinker of the Early Enlightenment, Bayle has been inserted by historians into both these overarching narratives.

Bayle is widely admired, of course, as one of the most sophisticated and passionate early advocates of toleration. In his *Commentaire philosophique* (1686)—his most direct and extensive engagement with that issue—he specifically included Judaism in his argument for the toleration of religious sects of all stripes (CP: OD II, 419). In his Bible criticism, though, Bayle focused with devastating relentlessness on the moral shortcomings of the Jewish heroes of the Old Testament. His witty but pointed *Dictionnaire* articles on biblical characters stand at the head of what developed into a barbed eighteenth-century tradition of irreligious polemic advanced through the ridiculing of Judaism (Sutcliffe 2003b). His 'David' article, which particularly outraged the Walloon Consistory of Rotterdam, was indeed a source for both Voltaire and d'Holbach, who reworked Bayle's arguments into their far more unambiguously anti-Judaic writings (Vercruysse 1978). The political and rhetorical legacy of Bayle with respect to Jews is, then, not clear-cut.

It is, though, intellectually fruitless to seek to categorize Bayle as either philosemitic or antisemitic, or as having been in essence positively or negatively disposed to Jews or Judaism. Rotterdam in the late seventeenth century hosted communities of both Ashkenazic and Sephardic Jews: in 1690 around 80 Jewish families in total,

---

[1] For an overview of the place of Judaism in Enlightenment thought, see Sutcliffe 2003a; for a sharply critical interpretation of this relationship, influential at the time of its publication though now largely discredited in scholarly debate, see Hertzberg 1968; on the conceptual and political history of 'Jewish Emancipation' see Sutcliffe 2018.

or approximately 400 individuals, resided there (Israel 2002, 96). It was predominantly a commercial centre, where, as in Amsterdam, the local Jews were disproportionately active and visible in the economic bustle that continually intersected with the cultural and intellectual life of the city (Van Lieshout 2008). Bayle, then, would certainly have encountered a broad variety of Jews in the course of his everyday life. There is no evidence, though, that these encounters made any particular impression on him. The lives of his Jewish fellow-immigrant neigbours do not seem to have stirred his curiosity. Bayle's extensive surviving correspondence, which fills 15 volumes in the recently completed scholarly edition published by the Voltaire Foundation, confirms the impression of him as a figure embedded above all in his own cultural network of the Huguenot diaspora. Elizabeth Labrousse's annotated list of his known correspondents is densely populated with Huguenot pastors, and does not include a single Jew (Labrousse 1961, 329–406). Bayle, we must conclude, was not actively interested in Jews or Judaism as a living reality in his own time. His was extremely interested, though, in the ways in which Jewish-related themes or examples could sharpen key philosophical arguments.

In Bayle's mind, most fundamentally, Jews occupied a uniquely anomalous position in relation to the philosophical question that preoccupied him throughout his life: the inescapable incommensurability, as he saw it, of faith and reason. Moral thinking, as presented in his writings, requires constant tacking between these two perspectives, which are each repeatedly shown to be insufficient on their own but irreconcilable with the other. Jews alone, as conceived by Bayle, stood outside this eternally inconclusive state of shuttling between reason and faith. As the recipients of God's directly revealed and legally authoritative Word, their religious and intellectual lives were uniquely unfractured. Reason and faith did not strain against each other for Jews, or at least they should not do so, because all aspects of Jewish life and law were directly subject to the authority of divine revelation.

An exploration of the place of Jewish themes in Bayle's work, then, while it certainly should form part of any survey of the wider mosaic of his cross-cultural curiosity and intellectual cosmopolitanism, also offers an exceptionally penetrating perspective on his wider philosophical thinking. This topic sharply elucidates his conception of the relationship between faith and reason, and especially of the role of conscience in navigating this. These issues stand at the heart of his arguments on toleration, and particularly valuable insights are therefore yielded by considering his approach to Jews and Judaism in the context of that key Baylean concern. The Jewish case, in summary, serves as a uniquely illuminating and challenging testing ground for the conceptual and political boundaries of Bayle's thought.

2.1 Bayle and the Old Testament

How interested was Bayle in Judaism? Richard Popkin rightly observed that, given the unique and unprecedented opportunities for interaction between Christians and Jews in the Dutch Republic in this period, it is notable that Bayle evinced no particular interest in theological discussion with living Jews. There is no trace in his writings, for example, of the 'friendly conversation' between the Remonstrant theologian Phillip van Limborch and the Sephardic doctor Orobio de Castro, van Limborch's account of which was published in 1687 in nearby Gouda (Popkin 1996, 635–6). Bayle was keenly interested, however, in Jewish history and in works of Hebraist scholarship. All of the Dutch francophone literary journals of the late seventeenth century gave prominent coverage to Hebraic subjects, but the *Nouvelles de la république des lettres* under Bayle's editorship (1684–7) paid them particularly close attention: there was, on average, at least one review of a Jewish-related book in each monthly issue (Sutcliffe 2003a, 33). An important conduit of knowledge for him in this area was his close friend Jacques Basnage, to whom he left most of his belongings in his will. Basnage, who may well have played a significant role in spurring Bayle's early interest in biblical criticism, was during Bayle's final years hard at work on his pioneering multi-volume *Histoire des Juifs*, which was published by Rainier Leers in Rotterdam in 1706, the year of Bayle's death (Bernier 2017, 245–6; Israel 2019, 481; Sutcliffe 2003a, 81–2).

Bayle's own facility with the Hebrew language was minimal, but he nonetheless grappled seriously with rabbinic texts, using Christian Hebraist reference sources as a guide. H. H. M. van Lieshout has counted 10 references in the *Dictionnaire* to Giulio Bartolocci's vast four-volume *Bibliotheca magna rabbinica* (1675–93), the most recent and comprehensive Hebraist project of the period (Van Lieshout 2001, 231). His attitude to rabbinic writings, however, was far from respectful. A critique of 'rabbinical fantasies' is a recurrent theme in the *Dictionnaire*, featuring frequently in the notes of his articles devoted to Old Testament figures, which mostly draw attention to the moral failings of those individuals. There are 21 such articles in the *Dictionnaire*, which constitute, according to Elizabeth Labrousse's calculations, 1% of the entries in the work (Labrousse 1964, 333). This is not a vast proportion, to be sure; but it is not negligible either, particularly when seen in the context of the intentional eclecticism of the *Dictionnaire*, which was one of the means through which Bayle conveyed his implicit argument that the boundaries of knowledge were inherently unfathomable and unruly. The contrast with Bayle's much lighter coverage of the New Testament, also noted by Labrousse, is certainly significant. Bayle devoted only two articles to New Testament topics (on John the Evangelist and the Sadducees). His biblical commentary was overwhelmingly focused on exploring the moral laxity displayed in the Jewish scriptures.

The typical pattern of Bayle's double-pronged critiques of the Old Testament and of rabbinic literature is established in the very first article of the *Dictionnaire*: on Aaron, 'high priest of the Jews, and brother of Moses'. In the third sentence of this article, Bayle offers this dry judgment on his subject: 'I would simply say that his weakness in lowering himself to the superstitious desires of the Israelites in the affair of the golden calf has given rise to many lies' (DHC, Aaron). In the remark here appended—the first remark of the work—Bayle elaborates on this by cataloguing the various 'rabbinical chimeras' that have been devised to exonerate or minimize Aaron's responsibility for this episode of idolatrous worship (DHC, Aaron, A). This article is paradigmatic of Bayle's biblical critique, the typical structure of which was to observe the moral failings of the ancient Hebrews and their revered leaders as recounted in Scripture, and then to expose the incoherence of later attempts, both by rabbis and by Christian theologians, to explain and excuse this behaviour.

The most famous and detailed example of this line of argument is the lengthy article on King David, which so scandalized his peers that Bayle bowed to the demands of the Walloon Consistory and cut the most provocative passages from the *Dictionnaire*'s second edition. Bayle here exposes with wit and relish the dramatic discord between David's great status and closeness to God, and his licentious, merciless and deceitful behaviour. He condemns with particular vigour David's conduct on the battlefield, which made the cruelty of modern warfare—even that of the Turks and the Tartars—seem mild in comparison (DHC, David, H).[2] In slaughtering the Amalekites following their raid on the town of Siceleg, Bayle tells us, David and his entourage 'did not let their swords rust … they killed men and women mercilessly, leaving only their cattle alive. … Frankly, this behaviour was very bad' (DHC, David, D).

Bayle's scrutiny of David's behaviour leads him to swiftly to the philosophical problem that lies at the heart of his engagement with Judaism and the Old Testament: the clash between reason and faith. This emerges here in the dissonance between the self-evident immorality, when judged on any rational basis, of David's behaviour, and the lofty status that he nonetheless retains for Christians, on the basis of faith in the divine provenance of the biblical record. In places Bayle seems to be attempting to mitigate this clash. On the slaughter near Sicileg, he comments that had David been acting under explicit divine orders it would not be appropriate to describe his behaviour as inhumane, but in this case it is clear that there was no such divine command, and that David himself was responsible for his actions (DHC, David, D). The underlying puzzle, though, remains. It is essential, Bayle argues, that morally wrong actions are condemned as such even when they are perpetrated by figures such as David who have been recipients of divine inspiration. There can be no middle path: either actions such as his are wrong, or similar actions committed by other people are not wrong. Attempts to argue otherwise are incoherent, and undermine the authority and clarity of moral standards. They thereby 'do great damage to the eternal laws, and consequently to true religion' (DHC, David, I).

Other articles intertwine a similar moral critique of the Old Testament with a more philosophically complicated probing of the dissonance between faith and reason, and of the uniquely anomalous status of Judaism in relation to this dissonance. The entry on Abimelech, for example—the Philistine king encountered by both Abraham and Isaac in the Book of Genesis—engages with the scriptural narrative on multiple levels. In the core article, as was typically his practice, Bayle's exegesis appears only mildly unconventional. Was it indeed the same Abimelech, he ruminates, to whom both Abraham and Isaac falsely presented their wives as their

---

[2] The lettering of remarks for this article follows the first edition (1697). In the expurgated version of the article that was included in subsequent editions, the lettering of the retained remarks was altered.

sisters? In the earlier incident, according to Genesis, Abimelech then took Sarah as his own wife, and would have deflowered her were it not for the miraculous intervention of God, who appeared to Abimelech in a dream and commanded him to restore her to her rightful husband Abraham. Bayle holds Abraham responsible for this narrow escape, noting that he could have avoided all risk if he had truthfully presented himself at the outset as Sarah's husband (DHC, Abimelech). In an early remark, he is scathingly critical of the interpretation that God had protected Sarah by afflicting Abimelech with a mysterious disease the rendered him impotent: this claim, according to Bayle, originated with Josephus and was reiterated by later rabbinical commentators. 'In am not surprised by the fantasies that the Jews have spouted over this episode', Bayle wrote, observing that he would have been more surprised if they hadn't 'created a hundred chimeras about our Abimelech' (DHC, Abimelech, B).

Attacks on the fantastical nature of rabbinic biblical commentary were part of the rhetorical stock-in-trade of seventeenth-century Christian Hebraism. Bayle frequently echoed these charges in entirely conventional terms. For him, though, these allegations were an early argumentative step in his reworking of biblical criticism as material for his philosophical exploration of the multidimensional tension between faith and reason. His next step in this process was typically to push reason as far as it can go as an alternative hermeneutical strategy. Abimelech's sexual restraint with Sarah, he suggests, was no miracle, but a simple instance of the diminution, with age, of the lusts of youth. Quoting a saying of 'young libertines', Bayle notes that 'a time comes when one is too wise' (DHC, Abimelech, D). Elsewhere, in a remark to his article on Sarah herself, he returns to this same issue, and offers another rational explanation: monarchs at the time, he comments, had so many wives that even very beautiful ones were seldom enjoyed by their husbands (DHC, Sarah, C).

In Bayle's hands biblical criticism here becomes a terrain for candid and even mildly bawdy discussion of the extent and limits of male sexual desire. This is one of the many aspects of what Mara van der Lugt has described as 'the obscenity of Pierre Bayle', which was among the main charges levelled at the *Dictionnaire* by the Walloon Consistory. Bayle's obsecenity, she shows, was intentionally provocative, and an important component of his wider campaign against euphemism, double standards, and all other forms of intellectual evasion and dishonesty (Van der Lugt 2018a). With respect to the Bible, this provocation takes on a further significance. Interpreting Scripture through a rational lens, Bayle normalizes the Jews and other Near Eastern peoples who populate its narrative, presenting them as subject to the same lusts as anybody else. As with the ethical failings of David, though, this sits awkwardly alongside the revered status of the Bible, according to Christian faith, as a divinely underwritten and uniquely authoritative record of ancient Jewish history. Bayle's mischievous sexual candour here also mischievously highlights the dissonance between critical and faith-based approaches to Scripture.

In his Abimelech article Bayle reinforces this point not through further analysis of the biblical text, but by pursuing his criticism of Josephus. In a remark on the relationship between sacred and profane history, he comments that he has long felt indignant towards this Roman-Jewish historian, and also towards those who had failed to note and criticize a fundamental contradiction between his life and work: 'a man, who openly professes Judaism, a religion founded on the divinity of Scripture, dares to recount matters differently from how one reads of them in Genesis; he changes, he adds, he supresses the circumstances; in a word, he sets himself in opposition to Moses, in such a way as one or other of them must be a false historian' (DHC, Abimelech, C). Bayle here charges Josephus with doing what no Jew should do: contradiction of the revealed word of God. Judaism is founded on the divine revelation of Scripture—and for a Jew to undermine this is to violate their covenant with God. As a historian, Bayle makes clear, Josephus behaved perfectly normally. All ancient historians, he observes, have taken similar liberties with their sources: 'they have stitched supplements to them, and, not finding the facts developed and embellished in them in accordance with their own imagination, they have extended them and adorned them as they pleased; we accept all this as history' (DHC, Abimelech, C). As a Jew, however, Bayle finds Josephus guilty, in using these normal historiographical methods, of violating the fundamental essence of Judaism.

Interpreters of Bayle inclined to see dissimulation as the keynote of his writing might perceive this accusation as self-evidently absurd, and therefore intended not to implicate Josephus as a 'bad Jew' but to undermine the religious starting-point that leads to this flawed and intolerant judgment. By demonstrating how traditional belief in the divine authority of the Bible leads to the unavoidable conclusion that Josephus cannot be accepted as a historian despite being a typical example of one, Bayle exposes—according to this interpretation—the ridiculousness that follows from that religious outlook. It is impossible to exclude this reading categorically, and indeed at some level is surely does form part of Bayle's intended implication. He clearly does suggest that there is something absurd about Josephus, or of any other Jew, not being eligible as a historian of their own people. It is unsustainable, though, to take this as Bayle's only or primary argument. His approach to the Bible, in this article and elsewhere, is too richly textured and intricate to be plausibly interpreted as simply one of veiled ridicule. It is unlikely that interpretive consensus on Bayle's religious orientation will break out among scholars any time soon. Careful attention to Bayle's handling of Jewish matters in passages such as this one, though, strongly bolsters the case for taking seriously his interest in the paradoxes of religious faith.

Bayle's cultural milieu throughout his life was French Protestant, except during his brief spell in Toulouse as short-lived convert to Catholicism in 1669–70. His loyalty to this community following his reconversion to Protestantism came at considerable personal cost: not only exile from France, but also, as Bayle saw it, the sequence of events that led to his beloved brother Jacob's imprisonment and death in 1685. These biographical facts are themselves sufficient reason, as Hubert Bost has argued, for interpreting his approach to the Bible, and to Jews and Judaism, in the context of the attitudes of his associates and allies in his Huguenot world (2012, 187–91). His close friend Jacques Basnage, as we have noted, was deeply interested in Jews and their past—but the conceptual framework of his *Histoire des Juifs* was conventionally Protestant in theological terms. Basnage chided those who had been gratuitously cruel towards Jews. He endorsed, however, the view that their exile and suffering was a divinely ordained punishment for their rejection of Jesus, and also believed in the inevitability and messianic importance of their ultimate conversion to Christianity (Basnage 1716; Sutcliffe 2003a, 82–9).

Bayle's thinking on the Bible was clearly far more subtle and philosophically probing that Basnage's. It is difficult to imagine, that he could have sustained such a close friendship with him, as well as very warm relations with many others who shared Basnage's conventional Calvinist outlook, if he did not sustain some respect for their committed faith in the Bible's textual authority. Bayle's painstaking engagement with the Bible, and with the problem posed by its apparent endorsement of unethical behaviour, also surely only makes sense if he believed that the authority of the Bible was not something that could be casually dispensed with. Jews, both within the biblical narrative itself and as later heirs of its most absolute authority, stood for Bayle as emblematically caught within the paradoxes generated by the incommensurable indispensability of both reason and faith.

2.2 Early Modern Jews in the Dictionnaire

Bayle's interest in Judaism was not restricted to the biblical era. Several articles in the *Dictionnaire* are devoted to Jews from the recent past, including some to near contemporaries. Ambiguities of identity fascinated Bayle, and he was particularly interested in the figure of the convert, perhaps because of his own personal experience of conversion both to and from Catholicism. He devoted one article to a possible convert from Judaism to Protestantism—the early seventeenth-century Hebraist Johann Rittangel—and another on a certain convert from Judaism to Catholicism: the sixteenth-century doctor, Paul Weidnerus, who ultimately became a professor of Hebrew at the University of Vienna, as did Rittangel in Königsberg (Popkin 1996; Van der Wall 1988). In a remark on the Wiednerus article, Bayle wrote with passion on the internal torment and externally inflicted suffering experienced by converts. Emphasizing converts' communality of experience, whether converting from Judaism to Christianity or between Christian denominations, he exhorted his readers not to forget 'a truly terrible type of persecution: of those who change their communion' (DHC, Weidnerus, A). As so often in the *Dictionnaire*, a philosophical provocation lay behind this point. The truth or falsehood of the confession joined by a convert, Bayle here unspokenly suggested, was irrelevant to the difficulties they were likely to experience, and also to the extent to which these challenges were deserving of understanding and sympathy.

The most subtle exploration of the complexities of conversion in the *Dictionnaire*, however, focused on a particularly tangled case of failed coming to terms with Judaism. This is found in the 'Acosta' entry, on Uriel da Costa: an early seventeenth-century Portuguese New Christian, who migrated to Amsterdam and joined the recently established Sephardic community there, but was expelled from it following his publication in 1623 of his *Exame das tradições phariseas*, an uncompromising critique of rabbinical authority (Costa 1993). Originally from Porto, Da Costa underwent circumcision in Amsterdam as part of the process of affiliating with his ancestral religious tradition, but it then took him only a few days, Bayle tells his readers, to realize that the customs and rituals observed by his fellow Jews did not conform to the Mosaic Law, and were enforced by rabbis who displayed 'neither heart nor pity' (DHC, Acosta). Bayle here clearly invites sympathy with Da Costa and his critical insights, which he sets in contrast to a stock Christian portrayal of spiritually sterile and intellectually dishonest rabbinism.

In his account of the rest of Da Costa's life, though, Bayle's focus shifts to the same theme as in his 'Weidnerus' article: the cultural isolation and mental anguish of the convert. Expelled from the Amsterdam Sephardic community, Da Costa (on Bayle's account) took refuge in Hamburg, where he was also ostracized by the local Sephardim. His religious opinions then further radicalized, to the extent of questioning the authority of the Mosaic law itself. Later, however, he recanted his opinions, and was subjected to a humiliating ritual of readmission to the Amsterdam community. Vividly evoking the mental agonies that led to this recantation, Bayle offers his readers his own imagined version of Da Costa's internal monologue at this juncture in his life, wrestling with the pain of isolation from his community and also, due to language barriers, from everybody else in his unfamiliar environment (DHC, Acosta).

In the remarks appended to this article, though, Bayle's probing of Da Costa's case leads beyond the emotionally wrenching but philosophically fairly straightforward terrain of the psychic perils of breaking with one's religious community. In general, as we have noted, Bayle considers this issue in ecumenical terms: these perils are the same regardless of the truth or falsehood of the religion an individual might be joining or leaving. With respect to Da Costa, though, Bayle ultimately pursues a different argument. This comes most clearly into focus in the article's final remark, in which Bayle comments on Da Costa's tragic death: shunned and despised even after abasing himself before the Amsterdam Sephardim, he committed suicide in 1640. Bayle here presents Da Costa's life as a vivid warning against pursuing philosophy too far, to the point of questioning revelation itself:

> Acosta [...] serves as an example. He did not want to acquiesce to the decisions of the Catholic Church, because he did not find them in accordance with reason; [...] next, he rejected a myriad of Jewish traditions, because he judged them to be unknown in Scripture; [...] finally, he denied the divinity of the books of Moses, because he judged that natural religion did not conform with the ordinances of this legislator. If he had lived another six or seven years, he would perhaps have denied natural religion, because his miserable reason would have found some difficulties with the hypothesis of providence and of the free will of an eternal and necessary Being [DHC, Acosta, G].

Linking Da Costa to his central philosophical preoccupation of the relationship between faith and reason, Bayle observes that there is nobody who can unerringly use reason without the need for assistance from God. Without this assistance, reason is 'a misleading guide'. Da Costa's life, for Bayle, eloquently exemplified the dangers of unrestrained rationalism, progressing, in his case, from a reasonable critique of rabbinism to the rejection of fundamental religious truths without which it was impossible for him to lead an intellectually coherent and emotionally happy existence. Bayle concludes this final remark by drawing this resonant moral from Da Costa's sad fate: 'Philosophy first refutes errors; but, if one does not stop it there, it attacks truths, and if one lets it do as it wishes, it goes so far that it loses its bearings, and can no longer find anywhere to stop' (DHC, Acosta, G).

Why does Bayle choose Uriel da Costa as his peg on which to hang this reflection on the dangers of untrammelled reason? The story was a recent and relatively local one, and it had recently been brought to renewed public attention in Bayle's 'Republic of Letters' by the publication of Costa's ostensible autobiography as an appendix to Van Limborch's account of his 'friendly conversation' with Orobio de Castro (Costa 1687). This volume was promptly reviewed in the *Histoire des ouvrages des savans*—the successor journal to Bayle's own *Nouvelles de la république des lettres*, edited by Jacques Basnage's brother, Henri Basnage de Beauval—and was Bayle's key source for this article. It would be wrong to conclude, though, that Da Costa caught Bayle's eye simply because his case was currently in the air, and that a similarly radical Christian deist might have served his philosophical purposes just as well. Da Costa's Jewishness enabled Bayle to suggest sharply distinct subversive arguments with respect to Christianity and to Judaism. The first step of Da Costa's intellectual journey—his rejection of the rabbis 'pharisaic' authority—would raise the hopeful expectation among Bayle's readership that he would then embrace the truth of Christianity, which was of course conceived as being founded on precisely this realization. In showing that Da Costa's reason did not lead him to this religious destination, Bayle's underscores his core argument that Christian belief is not grounded on reason, but on faith. Judaism, though, as conceived by Bayle and his Early Enlightenment contemporaries, was a religion not of reason or of faith, but of obedience to law. Da Costa's further radicalization, in which he lost his belief in the authority of that law, leaves him entirely lost. It is because Da Costa lacks the antidote to unchecked reason—faith, implicitly conceived here as intrinsically Christian, in contrast to Jewish legalism—that his fate is so tragic and disastrous.

The longest article in the *Dictionnaire* is devoted to an even more recent and much more renowned renegade from the Amsterdam Sephardic community. Spinoza was 8 years old at the time of Da Costa's suicide, and although his early influences remain unclear it can safely be assumed that he would have acquired detailed knowledge of the older heretic's arguments prior to his own communal expulsion in 1656. Bayle's 'Spinoza' article is also arguably the work's most influential, as reading between the lines of Bayle's extensive critique of Spinoza's arguments was one the most accessible ways in the early eighteenth century of gleaning a sense of what those dangerous arguments actually were. The paradox at the fore of this article was that of the 'virtuous atheist', of which Spinoza, both in his unblemished life and his calm and fearless death, was presented by Bayle as the supreme example. This popular image of Spinoza, which endures to this day, was diffused in the Enlightenment era above all by this article (Israel 2001, 298–301).

Bayle presents Spinoza's life as divided into three segments: he was 'a Jew by birth, and afterwards a deserter from Judaism, and lastly an atheist' (DHC, Spinoza).

Following his final embrace of atheism, Bayle seems to suggest, Spinoza's distance from Judaism was somehow categorically greater than it had been simply by virtue of his exit from his birth religion in 1656. The notion that the intellectually mature Spinoza was in no sense any longer a Jew was also present in the earliest biography of him, by Jean Maximilien Lucas, a Huguenot who had also been an admiring member of his circle. Very clearly for Lucas, and more subtly for Bayle (for whom Lucas was an important source), there was an echo between Spinoza and an earlier Jew who, it was believed, had rejected and transcended the narrow dogmas of his people. The image of Spinoza as the archetypical virtuous atheist became associated in the eighteenth century with the elevation of him by some to a status implicitly analogous to Jesus for the emerging age of reason: as, in effect, the messiah of the Enlightenment (Lucas 1927; Sutcliffe 2000, 2003a, 133–47).

Bayle's relationship to this image, though, despite having been such a key propagator of it, was complicated. Despite his admiration of Spinoza's character, Bayle's condemnation of his philosophy was, as Elizabeth Labrousse noted, exceptionally vehement by the standards of his normally playful and genial writing (Labrousse 1964, 198). The systematicity and unwavering certainty of Spinoza's marriage of atheism with geometrical reasoning appeared to Bayle as not only in some respects absurd but also profoundly hubristic and dogmatic. Bayle objected most fervently to Spinoza's rationalist dogmatism, which he perceived as the most fundamental and dangerous characteristic of the Spinozistic system of thought (Sutcliffe 2008, 70–76).

Geneviève Brykman is right, then, to note that in this article it is not Spinoza that Bayle puts on trial, but human reason (Brykman 1987–88, 269). Reason is also on trial in the 'Acosta' article, and it is by no means a coincidence that Bayle twice chooses to make this case with heretical Jewish exponents of reason as his foil. In both articles, and in Bayle's thinking more broadly, Judaism stands in antithetical contrast to reason. The two individuals with whom the articles are concerned are however conceived as standing in sharply contrasting relationships to their ancestral religion. Spinoza, whose uncompromising rebellion from Judaism, like that of Jesus, amounts to a radical inversion of the ancient religion's fundamental flaw, is in Bayle's eyes not simply an ex-Jew but an anti-Jew: the opposite of his former self. Da Costa, in contrast, did not make this exceptional leap. He remains, as understood by Bayle, still within the bounds of his Jewish particularity, and still subject to the unique regime of divine obedience that comes with it. From Bayle's perspective, this anomalous regime, which binds the Jews to God and to their divinely revealed law, is for Christians a mystery of faith. He highlights a similar mystery in his approach to the Old Testament: the immorality of David and others that is recounted and not condemned there must be in some sense confronted in its puzzling awkwardness, and not casuistically explained away. In contrast to Spinoza, whose systematic rationalism and atheism somehow flips him out of the Jewish orbit altogether, Da Costa's use of reason to critique the Mosaic law places him, alongside Josephus, in the category of 'bad Jew'.

2.3 Judaism, Toleration, and the Rights of Conscience

Bayle was fundamentally preoccupied by the issue of toleration. Some recent scholars, reacting against the earlier tendency to frame his thought around issues of reason and belief (or unbelief), have powerfully argued that we should regard toleration as his central underlying theme throughout his career, from his first writings in the early 1680s through to his late polemics against Jean le Clerc and his fellow *rationaux* in the years before his death in 1706 (Hickson 2013; García-Alonso 2019). The impact on him of both political and personal events in the autumn of 1685, as Labrousse and others have highlighted, were of transformative importance in imbuing this issue with urgency and tragedy. In November of that year, less than a month after Louis XIV's revocation of the Edict of Nantes, Pierre's beloved older brother Jacob died in French government custody in the Pyrenean town of Pamiers, where he had been held for over 5 months. Bayle was convinced that his brother had been held as a surrogate target for himself, and was therefore beset with a sense of responsibility for Jacob's death (Labrousse 1963, 198–9). This traumatic event was surely crucial in cementing in Bayle his profound sense of the unfathomability of any conceivable order of divine justice, or of any conceivable rationalization of evil: Bayle's own best efforts to speak truth to power had led only to intensified hardship and expulsion for his Huguenot peers in France, and to the death of his own brother. Jacob's death also entrenched Pierre's commitment to pacifism, which, we can surmise, he saw as the only viable strategy for avoiding such horrors. For Bayle, as Mara van der Lugt has aptly summarized, 'war … was the greatest of evils; peace, the highest good' (2016, 116).

This perspective set Bayle apart from the mainstream mood in the Huguenot diaspora, which was one of incensed anger toward the oppressiveness of Louis XIV's regime. The first extensive Huguenot exploration of the idea of toleration was Henri Basnage de Beauval's *Tolérance des religions* (1684), published in Rotterdam, perhaps in large part thanks to Bayle's efforts, while its author was still in France. The tone of this essay is for the most part vehemently polemical. Basnage de Beauval rails against the violent zeal of the French Catholic Church, against which he contrasts the stoic and honourable peacefulness of the Huguenot community (de Basnage 1970; Labrousse 1963, 216). Despite ostensibly advocating religious toleration, it is hard to see how, given his implacable fury towards Catholicism, the younger Basnage imagined that toleration might be extended beyond narrowly Protestant confines. Pierre Jurieu, meanwhile, responded to the Revocation with bellicose rage underwritten with a theologically conviction of righteousness. In his *L'accomplissement des prophéties* (1686–88), he argued the triumph of the true Church over the evil of popery was a biblically foretold inevitability. The intolerance of the French was abominable, for Jurieu, because it consitututed the persecution of truth by falsehood. Uncompromising intolerance of Catholicism, however, was not only politically justified but divinely commanded (Bracken 2001).

In his *Ce que c'est que la France toute catholique sous le règne de Louis le Grand* (1686), Bayle himself delicately juxtaposed, in letter form, two contrasting

Huguenot responses to the Revocation: polemical outrage and pragmatic moderation. It was not until his *Commentaire philosophique*, though, the first two parts of which appeared a few months later in the same year, that Bayle put forward his consolidated case for toleration, drawing together a range of ideas that he had explored in his earlier works (Laursen 1998, 204). He structured his argument around a refutation of the literal interpretation of Jesus' words in Luke's gospel: 'Go out into the highways and hedges, and compel them to come in, that my house may be filled' (Luke 14:23). Since Augustine's original exegesis, this prooftext had been widely invoked as a justification for coercing heretics back to the true faith. Strenuously repudiating this logic, Bayle here argues for the toleration of all sects, on the sole condition that they pose no threat to the state. He allows for no exceptions to this, and explicitly includes Jews among the religious communities that ought to be tolerated (CP: OD II, 419–20).

The key principle underpinning Bayle's argument in the *Commentaire*, as many commentators have noted, is the paramountcy of conscience (Kilkullen 1988, 59–105; Bost 1994, 51–62; Mori 1999, 273–304). He opens his text with a detailed defence of the necessity of basing all understanding on the inner insights of the individual 'natural light' (CP: OD II, 363). For Bayle, the inner voice of conscience is the voice of God, and therefore must in all circumstances be obeyed. The violation of this principle is so self-evident than it could only be accepted by someone bordering on insanity: 'as long as a man is not raving mad, he will not consent to anyone being able to command him to hate his God and to scorn his laws clearly signalled by conscience and intimately engraven in the heart' (CP: OD II, 384). In rebuttal of the argument, espoused most prominently by Jurieu, that a distinction must be made between the toleration of truth and of falsehood, Bayle insists that 'a conscience that is error has the same rights as one that is not' (CP: OD II, 422). The individual, he argues, is always morally bound to follow the dictates of their conscience, regardless of what these dictates might be, and whatever their religious basis.

This commitment leads Bayle to an immediate problem. What of those who believe that their conscience requires them to persecute others? This was in effect the argument of Louis XIV and his supporters, whose Catholic beliefs affirmed their sense that they were obliged to extirpate the heresy of Protestantism in France. Bayle does not flinch from this issue, and explicitly accepts this this troublesome conclusion: 'I do not deny that those who are actually persuaded that it is necessary to extirpate sects in order to obey God are obliged to follow the stirrings of this false conscience'. He adds, though, two important caveats. 'It does not follow that they what they do by conscience is not a crime', and moreover 'this does not mean that we do not have to speak out loudly against their false maxims and endeavour to enlighten their spirits' (CP: OD 430–31).

This shift of argumentative ground, far from representing the collapse of Bayle's case for toleration, as Walter Rex long ago influentially argued (Rex 1965, 184–5), in fact marks its philosophically crucial and characteristically Baylean inflection point. Anticipating his repeated highlighting in the *Dictionnaire*, as we have noted, of the ultimate insufficiency of human reason, Bayle here argues that reason cannot conclusively undergird the case for the principle that he felt was more important than all others: tolerance. Reason must here operate in awkwardly incommensurable tandem with faith. It is as an assertion of moral values, grounded in religious faith and therefore impervious to rationalist attack, that Bayle declares that religious persecution, even if carried out in accordance with personal conscience, is stained with 'crime' (he uses the French word 'crime', which has religious overtones of 'sin'), and therefore 'false'. In support of this swerve to the assertion of faith-based morals, he puts forward an analogous claim with respect to the regulation of sexuality. He does not need to be detained by the argument that someone might preach, in good conscience, in favour of sodomy and adultery, because this would never happen, and if anybody did 'preach sodomy' they would by met with such immediate public horror that they would never gain a band of followers (CP: OD 431). Bayle presumably made this digressive point in order to reinforce the alignment of his argument on toleration with his readers' broader moral intuitions. The distance from twentieth-first century secular norms of what be conceived as self-evidently sinful, though, highlights the extent to which his moral outlook was shaped by seventeenth-century Calvinism.

The significance of religion in Bayle's philosophy is underscored by his exceptionalist treatment in the *Commentaire* of the relationship of Jews to the dictates of conscience. He embarks on this topic in the very first chapter of the work, in which he stresses that the Old Testament falls within the purview of his argument. The Mosaic Law, he asserts, is fully compatible with the insights of the universal inner light, while also extending beyond it. For the Jews, both in Moses' time and since, these laws have served as a supplement to the inner light of conscience, dependent on it and as authoritative as it, just as deductions in geometry carry the same authority as the axioms from which they are derived (CP: OD II, 369). To elucidate this, Bayle discusses the case, discovered by him in the writings of the thirteenth-century French bishop and scholastic philosopher William of Auvergne, of a group of medieval Jews who renounced Judaism because they came to believe that the Mosaic Law was nothing more than an assemblage of useless or absurd precepts, unsupported by any divine authority. Bayle is sharply critical of these Jews' apostasy. He judges them guilty of disregarding both 'the incontestable proofs of divinity which God himself had entrusted through Moses' and the 'good reasons' for their ceremonial laws, given 'the character of the Jews and their penchant for idolatry' (CP: OD II, 370).

Despite, then, Bayle's general insistence on the absolute paramountcy of individual conscience, Bayle nonetheless unreservedly condemns these Jews who followed their consciences in choosing to abandon Judaism. In this case, the rights of conscience are overridden by the authority of the Mosaic revelation, which, Bayle reminds his readers later in the *Commentaire*, is 'a general law for the Jews, enunciated absolutely and without restrictions of time or place' (CP: OD II, 408). Whereas for Christians the key source of divine and ethical knowledge is the personal inner light of conscience, for Jews this is trumped by the divinely revealed Mosaic Law, which placed the Jews under a 'theocratic government', which abrogated 'the immunities of conscience' (CP: OD II, 408).

The unique nature of God's direct revelation through Moses functions in the *Commentaire*, as in the later *Dictionnaire*, as the key and inescapable reminder of the insoluble mysteries of both faith and reason. God's suppression of the rights of conscience for the Jews, as part of the divine intimacy of God with this chosen people, constitutes in Bayle's eyes the most vivid possible testimony to the unfathomability of faith and of its relationship to reason. For from seeking to occlude this problem, he confronts it directly, and pushes it to its most challenging limit. Noting that the Mosaic Law allowed absolutely no toleration for idolators or false prophets, was it not also possible, he rhetorically asked, that the God of the New Testament might have commanded a similar intolerance towards heresy? Bayle's response to this argument is blunt and unflinching: 'I avow in good faith that this is a powerful objection, and seems to be a sign that God wants us to know almost nothing with certainty, through the exceptions He has placed in Scripture for almost all the common notions of reason' (CP: OD II, 407).

Bayle's deepest intellectual commitment was to the avoidance of any form of dogmatism. Religious arguments over rigid tenets of doctrine, he was convinced, could only lead to entrenched conflict and violence rather than to peace. His antidogmatic argument for toleration, though, in no sense disarmed him in his battle against his adversaries, among whom Jurieu, and his increasingly strident anti-Catholic dogmatism, was at this time moving to the fore. The clear commitment of Bayle's own conscience to intellectual openness, pragmatism and peace rather than fanaticism and war underpinned his passionate critique of Jurieu's worldview across the remaining 20 years of his life. In the *Dictionnaire*, he railed passionately against several historical figures, such as Savonarola and Comenius, whom he cast as 'fanatics' and would have readily identified by his readers as surrogate targets for Jurieu (Laursen 1998, 206–13; Van der Lugt 2016, 145–8). More recent readers, often eager to heroize Bayle as an early herald of inclusive liberal toleration and as an enduring valuable contributor to the philosophical arsenal against modern forms of fanaticism, have mostly ignored the exclusion from his conscience-based argument for toleration of Jews, whose political inclusion has since the late eighteenth century been widely seen as a fundamental hallmark of moral decency.

Bayle's belief that Judaism was incompatible with individual conscience—which he goes out of his way to highlight in the *Commentaire*, and also reiterates elsewhere, such as (in slightly different terms) in his comments discussed above on Josephus and on Uriel da Costa in the *Dictionnaire*—highlight two important aspects of his broader thinking. Firstly, we are reminded that Bayle, like most Enlightenment thinkers before the 1780s, did not give Jews serious consideration in relation to issues of practical politics, but tended instead to contemplate them as an essentially timeless conceptual thinking tool, tightly associated with the Bible and often figured as a foil to reason (Sutcliffe 2003a, 193–261; Sutcliffe 2020, 66–80). Secondly, the evidence that Bayle placed on the anomalousness of the Jewish case highlights the importance to him of signalling the central importance of the mysterious incommensurability of faith and reason. Twenty-first-century readers readily overlook these passages, because the interpretation of the Bible counts for little for most of us. For Bayle's contemporaries, in contrast, the question of how to read the

Bible was existentially crucial, and it was through his treatment of the Jewish case that Bayle wove that question into his wider hermeneutical and political arguments. If we read his work through seventeenth-century eyes, with close attention to his handling of the Old Testament and its Jewish custodians (as Bayle saw them), we more clearly perceive how Bayle's paradoxes of faith and reason bound his approach to toleration into his broader philosophy.

2.4 Late Bayle: Conscience in Conflict

Bayle's *Commentaire*—which he published anonymously, and never publicly acknowledged as his own—was soon attacked from all sides. Jurieu denounced the work as an anti-Calvinist denial of true religion, and firmly rejected the concept of freedom of conscience. Less fiery but argumentatively similar attacks came from Jurieu's more moderate opponents, such as Élie Saurin and Jean Le Clerc, who concurred that Bayle's position implied an indifference towards religious truth, and was therefore profoundly dangerous (Mori 1999, 286–95). Bayle now found himself in the position in which he would remain for the rest of his life: of being bitterly critiqued from two opposing sides.

This sense of intellectual isolation was greatly compounded by the Glorious Revolution of 1688, which Bayle, almost alone in Huguenot circles, bitterly opposed. For Jurieu, who eagerly assumed the role of house theologian to William of Orange's invasion of Britain, this hugely portentous event, nipping in the bud the emergence of a Catholic centre of power in London, presaged the impending triumph of Protestantism over Louis XIV's France. Bayle was horrified by this prophetic warmongering, and by the blurring of theocratic partisanship with politics that underpinned it. He saw the overthrow of James II as an act of political sedition. James, he noted, had not required Protestants in Britain to abandon their faith. On the contrary: he had committed to a policy that defended freedom of conscience. The Protestant backers of the Glorious Revolution had refused to accept this, preferring instead to restore their own supremacy (García-Alonso 2019, 808–9). Bayle's non-partisan analysis of the religious politics of the Glorious Revolution was rejected not only by Jurieu and his camp, who openly embraced the Revolution's denominational partisanship, but also by Huguenot moderates, who sought to rationalize it away. Intellectuals such as Jean Le Clerc were closely allied with John Locke, whose hugely influential writings played a key role in presenting the Glorious Revolution as a triumph not of Protestant factionalism but of objectively reasonable good sense.

Locke's writings on toleration, which lent philosophical support to the Toleration Act passed by the English Parliament very soon after the Revolution, were central to this philosophical makeover. Scholarly attention has understandably focused on Locke's arguments for the political toleration of dissenting Protestants, but not of Catholics—but the unstable foundations of Locke's thinking on this issue are most sharply brought into relief by his handling of the question of toleration for Jews. In the first few lines of his *Letter concerning toleration* Locke frames his topic in explicitly Christian terms, as 'the mutual toleration of Christians in the different professions of religion' (Locke 1991, 14). Later, though, almost in passing, he declares himself in favour of extending toleration universally, to both Muslims and Jews. He defends this, though, on Christian theological rather than reasoned philosophical grounds: 'if we may openly speak the truth … neither Pagan, nor Mahometan, nor Jew, ought to be excluded from the commonwealth because of his religion. The gospel commands no such thing' (Locke 1991, 51). Swiftly challenged on this specific point by Jonas Proast, Locke shifted his ground. In his response to Proast, he reiterated his support for the toleration of Jewish worship, but defended this for conversionist reasons entirely unrelated to his wider theorization of toleration. It is a Christian duty, he argued, to pray every day for the conversion of Jews to Christianity, but this cannot by achieved by driving Jews away or persecuting them. The casual slipperiness of Locke's arguments on Jews reveals how marginal this question was to his more concrete thinking on intra-Christian toleration. It also highlights a more fundamental instability in the relationship between Locke's philosophical and theological thinking. The political issue of the toleration of Jews sat at the boundary between those domains, and Locke was unable to situate it clearly in either of them (Sutcliffe 1999, 2003a, 218–20).

Like Bayle, Locke grounded his argument for toleration on the primacy of conscience, which occupied a similarly foundational position in the philosophical and religious outlook of both men. Locke also shared Bayle's belief that Judaism stood uniquely outside the otherwise universal reach of the rights of conscience. Again like Bayle, he considered it necessary to state this clearly in the context of enunciating his theory of toleration. 'The commonwealth of the Jews, different in that from all the others, was an absolute theocracy', he declared, explaining that the Mosaic law, as divinely authored civil legislation, was absolutely binding for the people to whom it applied (Locke 1991, 39). Locke made this point in order to demarcate the boundary between his conscience-based theory of toleration and a theological domain where it was off limits. The religious world of the Old Testament—and implicitly also Judaism *in toto*, understood as a fossilized survivor from that era—was ceded by Locke to the jurisdiction of religion. The domain of Locke's liberal philosophy of individual rights, underwritten by the primary of conscience, was thus saved from the hubris of appearing to displace entirely the idea of supreme divine authority over human affairs.

Unlike Locke, Bayle did not believe that any such tidy demarcation was possible. Christian faith, even though sharply distinct from the Judaism of the Old Testament, nonetheless rested on a continued faith in the voice of divine authority recorded in the Jewish scriptures. Human reason, meanwhile, could not free humanity from the need for faith, as he had vividly demonstrated by showing, in the *Commentaire*, how appealing to the rights of conscience could not provide the basis for a watertight case for toleration. Bayle's turn, in the 1690s, to his *Dictionnaire* project was motivated by a desire not simply to explore this fundamental intellectual puzzle, but also, at a more important moral level, to challenge the evasiveness of those who thought it could somehow be sidestepped or ignored, and the arrogance of those who thought it could be transcended. It was precisely this arrogance that Bayle saw in both of the rival intellectual camps of the Huguenot diaspora: Jurieu and the religious fundamentalists, who asserted the absolute authority of faith, and Le Clerc and the *rationaux*, who affirmed that faith could be fully resolved with reason. Bayle's repeated use of Jewish examples in the *Dictionnaire* to highlight the the clash between faith and reason was one of the means through which he simultaneously challenged both these camps.

In Bayle's final works—his *Continuation des pensées diverses sur la comète* (1704), the four-part *Réponse aux questions d'un provincial* (1703–1707), and the posthumously published *Entretiens de Maxime et de Thémiste* (1707)—this critique moved centre stage. These are his most adversarial writings, directed in particular against Le Clerc and his fellow *rationaux*, whose arguments, he had come to believe, were ultimately no less dangerously dogmatic than Jurieu's explicitly intolerant religious zealotry. Bayle's biting attacks on Le Clerc's rationalist theology were in response to the intellectual intolerance of the *rationaux* towards Bayle's most provocatively heterodox arguments, on topics such as Manicheanism, in the *Dictionnaire*. Le Clerc, Bayle pointed out, not only deemed intolerable not only those arguments, but also by implication any argument that failed to recognize the harmony of religion and reason. This amounted to a refusal to tolerate almost everybody (EMT: OD IV, 31). In these late works, Bayle relentlessly sought to expose the intolerance and dogmatism of his adversaries, and thus to demonstrate the necessity of his own alternative: unflinching intellectual honesty and humility.

Recent Baylean scholarly debate has revolved around these late works, from which it is tempting to seek to distil a definitive argumentative climax that offers a clearly defined 'useable Bayle' for our own times. Michael Hickson has argued, for example, that Bayle's final work, the *Entretiens*, provides the key to his resolution of the problem of evil, achieved by conceptualizing toleration itself as a theodicy (Hickson 2010, 560–67). Gianluca Mori, in contrast, has interpreted Bayle as becoming less interested in religious pluralism and moving closer to explicit non-belief, presenting with great intellectual sympathy, in the *Continuation*, the contours of a rationalist, eternally questioning 'Stratonist' atheism (Mori 1999, 217–36). Neither of these interpretations, though, account for the anomalous place of the Old Testament and of Judaism in his thought. Bayle's exploration of Stratonism, and also his experimentation with the dialogue form through the voices of Maxime and Thémiste, were above all attempts to push the boundaries of non-dogmatic thinking. Reading these late texts as reaching towards a final philosophical resolution of the issues that seem so resolutely undecidable in his earlier writings obscures the central importance of paradox and undecidability throughout his work. The importance of undecidability for Bayle cannot, because of its very nature, be decisively clinched: as he observes in his *Dictionnaire* article on Pyrrhonistic scepticism, doubtful questioning is essential, but this also requires us 'to doubt if it is necessary to doubt' (DHC, Pyrrho, C). The persistent exceptionalism of Jews and Judaism in Bayle's writings nonetheless serves as a stubborn reminder of his philosophical resistance to closure.

Recent commentators have also made much of apparent shifts in Bayle's late work in his thinking on conscience. According to Mori and Anthony McKenna, in his final writings Bayle no longer argued for the limitless rights of the erring conscience, but emphasized rather the self-evidence of the fundamental and universal truths of moral rationalism (Mori 1999, 305–18; McKenna 2012, 93–98). Hickson, meanwhile, argues that in his late writings Bayle finally completes his intellectual journey towards a full 'secularization of conscience' (Hickson 2018). Several of his statements in these late works—such as his observation in the *Réponse* that 'a Spinozist king', or in other words a religiously non-aligned and possibly even atheist ruler, would be more likely than any Christian ruler to provide protection from partisan dogmatism and intolerance—certainly seem at first sight to suggest a turn towards rationalism (RQP: OD III, 954–5, 1011–12). Such interpretations, though, fail to take account of the charged polemical context of Bayle's writing after 1700. Bayle was engaged in an intellectual battle against adversaries who used some of the same conceptual resources as him, not in support of toleration but—on Bayle's interpretation at least—against it. These contested resources included the notion of conscience itself, which, as we have noted, John Locke (and also his *rationaux* allies) marshalled in justification of the hypocritically selective pseudo-toleration—again, as Bayle saw it—of the Glorious Revolution. Against this backdrop, and holding in mind the characteristic mobility and inconclusivity of so much of his writing and also his sharp critique of Spinoza, it seems more plausible to read Bayle's suggestion that Spinozistic atheism might be politically preferable to any kind of Christian rule not as some sort of final political testament, but rather as yet another provocative paradox.

A central theme of the third part of the *Réponse* is the barbarism frequently committed in the name of 'conscience'. It is in contrast to this, Bayle suggests, that the regulation of the public domain by a secular authority unswayed by appeals to religious conscience—such as a 'Spinozist king'—would be preferable. Among those who do terrible things in the name of conscience, he cites the example of Jews, who, he alleges, 'according to a principle of conscience only show the right path, or the way to the fountain, to those of their own religion, and consider every harm they can inflict on Christians as a good deed' (RQP: OD III, 986). This is a strikingly illiberal sweeping statement about Jews, to say the least, and one that cannot easily be reconciled with the presentation of Bayle as a straightforwardly progressive foundational thinker of 'Radical Enlightenment'. Jews here clearly occupy a position in Bayle's thought that is emblematic of a severely distorted conception of conscience. It is also clear that he is here using the word conscience in a very different sense than in the *Commentaire philosophique*, where conscience refers to the inner light of ethical judgment, which for Jews and Jews alone was overwritten by the unquestionable divine authority of the Mosaic law. That argument is readily forgotten by twenty-first-century readers of Bayle, but in his own milieu it was a commonplace, and an area of common ground between him and his *rationaux* adversaries. His description of Jewish obstructionism towards Christians as justified by conscience would have been immediately perceived by Le Clerc and his allies as a sharply ironic highlighting of how distorted this concept could become.

By the time of his final writings, then, Bayle regarded the term 'conscience' as so widely abused as to be almost valueless. There is no sign, though, that he abandoned his belief in a human 'inner light' of wisdom and decency, equally beyond religion and beyond rational logic. This is what, in the *Commentaire*, he meant by true conscience (as opposed to false or erring conscience), and appealing to it remained fundamental to his case for toleration. Bayle was open, in his non-dogmatic and pragmatic intellectual fashion, to conceiving of this inner light in varied ways, including as moral rationalism; the limited efficacy of that persuasive strategy was, though, glaringly obvious to him, as the moral rationalism of the *rationaux* in his view led straight back to intolerance. A faith-based understanding of true conscience, though, rested on similarly mysterious and insecure foundations. The textual container of those mysteries was the Old Testament, and the human reminders of them, down to Bayle's own time, were the Jews.

2.5 Conclusion: Bayle, Judaism and the Enlightenment

Bayle considers toleration in later writings, as a number of scholars in varying ways have noted, most importantly as a practical—or, to put it another way, political—problem, rather than as an issue that is soluble in theological or philosophical terms (Van der Lugt 2016, 243; García-Alonso 2017, 2021, 97). His demonstration in the *Commentaire philosophique* that appeal to the rights of the erring conscience did not provide a solid basis for universal opposition to toleration showed that the search for such a conclusive intellectual solution ultimately led to an impasse. The urgency of finding a way forward, though, was all the more keenly felt by Bayle, though, when he himself faced a double-pronged onslaught of intolerance from religious dogmatists and rationalists simultaneously. One way forward was to insist on the priority of actions over words. This, as Van der Lugt has shown, was an important argument lying behind Bayle's 'obscenity': censorship and euphemism around sex, he believed, were a hypocritical distraction from what really mattered in this domain, which was moral behaviour (2018a, 725–33). Bayle argued similarly with respect to toleration. His most explicit articulation of this argument, it is significant to note, was through the life of a Jew: the fifteenth-century Sephardic philosopher Isaac Abrabanel.

Abrabanel's great fault, Bayle observed at end of the *Dictionnaire* article on him, was that he was 'too sensitive' to the persecutions that the Jews, including Abrabanel himself, had suffered. These historical memories animated him with such fury towards Christians that 'he barely wrote a single book which was not stamped by his desire for vengeance'. His behaviour, however, was very different: 'the implacable hatred that he expressed towards Christians while writing did not prevent him from living with them in a civil, pleasant and gracious manner' (DHC, Abrabanel). Bayle's criticism to Abrabanel's excessive sensitivity is surely ironic, intended rather to signal how understandable his hostility to Christians was, given the experiences of Jews in the early decades of the Spanish Inquisition. Simultaneously, however, he portrays Abrabanel as standing outside the ethical norms of Christian culture. Forgiveness—a paramount ethical value in Christianity, and fundamental to Bayle's hope that a pathway to the mutual toleration of French Protestants and Catholics might be found despite the long history of violence between them—has no place in Abrabanel's fiery anti-Christian rhetoric. Inexplicably but thankfully, though, in his actual dealings with Christians he was exemplary.

Jews, then, could serve for Bayle as a positive model. To conclude from this that he was an admirer of Jews, though, would be to miss the point of his argumentative logic. Abrabanel was such a compelling example of generous behaviour contradicting hard-hearted beliefs precisely because, for Bayle, Jews were quintessentially representative of rigid obedience to the Mosaic Law. His cordiality towards Christians is therefore strikingly paradoxical, and reinforces the wider paradoxes of faith and reason that pervade the *Dictionnaire*. There is nothing in any way resembling a voice of conscience that leads Abravanel to behave decently towards his Christian neighbours. Bayle does not spell this point out, because he does not need to: the conception of Judaism as a religion in which there is no space for conscience is, as we have seen, a fundamental assumption not only of Bayle himself but also of his wider intellectual milieu. Abravanel's behaviour towards Christians is, precisely because of this, a choice example of the baffling mysteries of human existence with which Bayle was so preoccupied and fascinated.

The idea that Judaism was inherently and fundamentally impervious to the ethical framework on which toleration was based remained a prominent outlook among leading figures of the eighteenth-century Enlightenment. Voltaire, the most famous apostle of toleration in eighteenth-century France, repeatedly portrayed Jewish legalism and barbarism as the antithesis of his own positive values (Sutcliffe 1998, 2003a, 231–46). Gotthold Ephraim Lessing, who has the equivalent status for eighteenth-century Germany, seems at first sight to be very different: the eponymous hero of his play *Nathan der Weise* (1779), his most famous dramatic plea for toleration, was famously based on his close friend, the leading Jewish intellectual Moses Mendelssohn. Nathan, however, espouses a brand of universalism that is not only denuded of any Jewish content but is rather the antithesis of the pervasive eighteenth-century Enlightenment conception of Judaism. Like Spinoza for Bayle, Nathan and perhaps also Mendelssohn for Lessing were of interest not at Jews but rather as transformed post-Jews (Sutcliffe 2013).

Bayle's exceptionalist approach to Judaism does not, then, set him apart. Although the intricate argumentative ends to which he puts this exceptionalism are extremely distinctive, the Bible-based assumptions that underlie his thinking on Jews and Judaism were broadly shared by most leading European intellectuals of the Enlightenment era (Sutcliffe 2020, 62–106). A turn towards sincerity, and above all sincerity towards one's own conscience, was central to the ethical thinking of this period. Mara van der Lugt has pithily summarized this as a shift from 'truth' to 'truthfulness', which, she shows, links Bayle at the beginning of the Enlightenment with Kant at its end (2018b). For Kant, as for Bayle, this ethical primacy of autonomous truthfulness was the very antithesis of Jewish legalism. The triumph of 'pure moral religion', and its embrace by Jews, would amount, Kant famously declared, to 'the euthanasia of Judaism' (Kant 1992, 95).

We live today in an age of renewed ethical subjectivism, in which appeals to 'my truth' abound in the public sphere. Bayle's late writings on conscience, in which he is so keenly aware of how slippery this notion could be, can perhaps here offer valuable insights for our own times. His exceptionalist thinking on Jews and Judaism in this regard, though, sits very uncomfortably in the political and cultural context of the twenty-first century. The view that individual conscience is an ethical foundation intrinsically absent from Judaism would today generally be considered untenable. One important reason for this is that over the past two centuries, since the emergence of the Reform tradition in early nineteenth-century Germany, Judaism has itself become an extremely internally contested and varied religion. The intellectual heirs of Da Costa—intrinsically 'bad Jews' according to Bayle's logic, and also in the eyes of many Orthodox Jews—now have their own synagogues, espousing new and diverse interpretations of the Mosaic Law, rabbinic authority, and the relationship of the individual to the religious community and to the state.

The belief that the Jews, as 'God's chosen people', are of exceptional importance and require exceptional consideration, continues, though, to play a very important role in international politics. Evangelical Christian Zionists, through organizations such as the five-million-strong Christians United for Israel, exert significant influence on the Middle East policy of the United States and several other countries. This form of biblicist exceptionalism is closest in spirit, in Bayle's day, to the prophetic millenarianism of Jurieu. Controversies over the exceptionalist treatment of Jewish-related issues also figure prominently, however, at the interface of Western liberal moralism and those cultural systems—especially Islam—which are most commonly cast as its rhetorical adversary. This can be seen in arguments over the nature and seriousness of antisemitism in relation to Islamophobia and other prejudices, and over the pedagogical prioritization and political contestation of Holocaust memory. This clash was also in evidence at the 2022 FIFA World Cup in Qatar, where Western efforts to foreground gay rights were pitted against a predominantly Arab focus on Palestinian rights, which, these expressions of solidarity indirectly suggested, were largely 'beyond the conscience' of the West for reasons closely connected with deep-seated exceptionalist thinking on Jews (Sutcliffe 2020, 274–83; Sutcliffe 2022).

Bayle certainly does not offer us any easy untangling of these complex theologico-political knots. The insight that he offers, however, as several scholars have observed, does not lie in his conclusions, but in the anti-dogmatic open-endedness and undecidability of his method (Lennon 1999, 183–6; Whelan 2010; Russo 2019). His indefatigable grappling with the paradoxes and knots that most concerned him, and his absolute refusal of any evasions or false solutions of them, might, then, offer some sort of model for how we should approach these abiding cross-cultural muddles of our own time.

Bibliography[3]

Basnage, Jacques. 1716. *Histoire des Juifs, depuis Jésus-Christ jusqu'à present, pour servir de supplément et de continuation de l'histoire de Joseph.* The Hague.
Bernier, Jean. 2017. Pierre Bayle and biblical criticism. In *Scriptural authority and biblical criticism in the Dutch golden age: God's word questioned*, ed. Dirk van Miert, Henk Nellen, Piet Steenbakkers, and Jetze Touber, 240–256. Oxford: Oxford University Press.
Bianchi, Lorenzo. 1994. Bayle e l'ebraismo. In *La questione ebraica dall'illuminismo all'impero*, ed. Paolo Alatri and Silvia Grassi, 17–36. Naples: Edizioni Scientifiche Italiane.
Bost, Hubert. 1994. *Pierre Bayle et la religion.* Paris: Presses Universitaires de France.
———. 2012. Pierre Bayle: a 'complicated Protestant'. In *The persistence of the sacred in modern thought*, ed. Chris L. Firestone and Nathan A. Jacobs, 185–208. South Bend: University of Notre Dame Press.
Bracken, Harry M. 2001. Pierre Jurieu: the politics of prophecy. In *Continental millenarians: Protestants, catholics, heretics*, ed. John Christian Laursen and Richard H. Popkin, 85–94. Dordrecht: Kluwer.
Brykman, Geneviève. 1987–1988. Bayle's case for Spinoza. *Proceedings of the Aristotelian Society* 88: 259–270.
Costa, Uriel da. 1687. *Exemplar humanae vitae.* In *De veritate religionis christianae. Amica collatio cum erudito judaeo*, ed. Phillip van Limborch. Gouda.
———. 1993 [1623]. *Examination of pharisaic traditions*, eds. H. P. Salomon and I. S. D. Sassoon. Leiden: Brill.
Basnage de Beauval, Henri. 1970 [1684]. *Tolérance des religions.* Ed. Elizabeth Labrousse. New York: Johnson.
García-Alonso, Marta. 2017. Bayle's political doctrine: a proposal to articulate tolerance and sovereignty. *History of European Ideas* 43: 331–344.
———. 2019. Tolerance and Religious Pluralism in Bayle. *History of European Ideas* 45: 803–816.
———. 2021. Persian theology and the checkmate of Christian theology: Bayle and the problem of evil. In *Persia and the Enlightenment*, ed. Cyrus Masroori, Whitney Mannies, and John Christian Laursen, 75–100. Liverpool: Voltaire Foundation/Liverpool University Press.
Hertzberg, Arthur. 1968. *The French Enlightenment and the Jews: the origins of modern antisemitism.* New York: Columbia University Press.
Hickson, Michael W. 2010. The message of Bayle's last title: providence and toleration in the *Entretiens de Maxime et de Thémiste. Journal of the History of Ideas* 71: 547–567.
———. 2013. Theodicy and toleration in Bayle's *Dictionary. Journal of the History of Philosophy* 51: 49–73.
———. 2018. Pierre Bayle and the secularization of conscience. *Journal of the History of Ideas* 79: 199–220.
Israel, Jonathan I. 2001. *Radical Enlightenment: philosophy and the making of modernity 1650–1750.* Oxford: Oxford University Press.
———. 2002. The Republic of the United Netherlands until about 1750: demography and economic activity. In *The History of the Jews in the Netherlands*, ed. J.C.H. Blom, R.G. Fuks-Mansfeld, and I. Schöffer, 85–115. Oxford: Littman.
———. 2019. Pierre Bayle's correspondence and its significance for the history of ideas. *Journal of the History of Ideas* 80: 479–500.
Kant, Immanuel. 1992 [1798]. *The conflict of the faculties.* Trans. Mary J. Gregor. Lincoln: University of Nebraska Press.
Kilkullen, John. 1988. *Sincerity and truth: essays on Arnauld, Bayle and Toleration.* Oxford: Oxford University Press.
Labrousse, Elizabeth. 1961. *Inventaire critique de la correspondance de Pierre Bayle.* Paris: Vrin.

---

[3] NB: All references to DHC are to the first edition (1697).

———. 1963. *Pierre Bayle – I: Du pays de Foix à la cité d'Erasme*. The Hague: Martinus Nijhoff.
———. 1964. *Pierre Bayle – II: Hétérodoxie et rigorisme*. The Hague: Martinus Nijhoff.
Laursen, John Christian. 1998. Baylean liberalism: tolerance Requires nontolerance. In *Beyond the persecuting society: religious toleration before the Enlightenment*, ed. John Christian Laursen and Cary J. Nederman, 197–215. Philadelphia: University of Pennsylvania Press.
Lennon, Thomas M. 1999. *Reading Bayle*. Toronto: University of Toronto Press.
Locke, John. 1991 [1689]. *Letter concerning toleration*. Trans. William Popple. In *John Locke—A letter concerning toleration in focus*. London: Routledge.
Lucas, Jean Maximilien. 1927 [c. 1677]. The life of the late Mr. de Spinoza. In *The oldest biography of Spinoza*, ed. and trans. A. Wolf. London: George Allen and Unwin.
McKenna, Antony. 2012. Pierre Bayle: free thought and freedom of conscience. *Reformation and Renaissance Review* 14: 85–100.
Mori, Gianluca. 1999. *Bayle philosophe*. Paris: Champion.
Popkin, Richard H. 1991. In *"Introduction" to Pierre Bayle—Historical and critical dictionary: selections*, ed. Popkin, viii–xxix. Indianapolis: Hackett.
———. 1996. Pierre Bayle and the conversion of the Jews. In *De l'Humanisme aux Lumières: Bayle et le protestantisme*, ed. Michelle Magdelaine et al., 635–643. Oxford: Voltaire Foundation.
Rex, Walter. 1965. *Essays on Pierre Bayle and religious controversy*. The Hague: Martinus Nijhoff.
Russo, Elena. 2019. How to handle the intolerant: the education of Pierre Bayle. In *Imagining religious toleration: a literary history of an idea*, ed. Alison Conway and David Alvarez, 119–135. Toronto: University of Toronto Press.
Sutcliffe, Adam. 1998. Myth, origins, identity: Voltaire, the Jews and the Enlightenment notion of toleration. *The Eighteenth Century: Theory and Interpretation* 38: 67–87.
———. 1999. Enlightenment and exclusion: Judaism and toleration in Spinoza, Locke and Bayle. *Jewish Culture and History* 2: 26–43.
———. 2000. The spirit of Spinoza and the Enlightenment image of the pure philosopher. In *Heroic reputations and exemplary lives*, ed. Geoffrey Cubbitt and Allen Warren, 40–56. Manchester: Manchester University Press.
———. 2003a. *Judaism and Enlightenment*. Cambridge: Cambridge University Press.
———. 2003b. Judaism in the anti-religious thought of the clandestine French early Enlightenment. *Journal of the History of Ideas* 64: 97–117.
———. 2008. Spinoza, Bayle, and the Enlightenment politics of philosophical certainty. *History of European Ideas* 34: 66–76.
———. 2013. Lessing and toleration. In *Lessing and the German Enlightenment*, ed. Ritchie Robertson, 205–225. Oxford: Voltaire Foundation.
———. 2018. Toleration, integration, regeneration, and reform: rethinking the roots and routes of 'Jewish emancipation'. In *The Cambridge history of Judaism, volume VII: the early modern world, 1500–1815*, ed. Jonathan Karp and Adam Sutcliffe, 1058–1088. Cambridge: Cambridge University Press.
———. 2020. *What are Jews for? History, peoplehood, and purpose*. Princeton: Princeton University Press.
———. 2022. Whose feelings matter? Holocaust memory, empathy, and redemptive anti-antisemitism. *Journal of Genocide Research (online first)*.
van der Lugt, Mara. 2016. *Bayle, Jurieu, and the Dictionnaire historique et critique*. Oxford: Oxford University Press.
———. 2018a. Les mots et les choses: the obscenity of Pierre Bayle. *Modern Language Review* 113: 714–741.
———. 2018b. The left hand of the Enlightenment: truth, error and integrity in Bayle and Kant. *History of European Ideas* 44: 277–291.
van der Wall, Ernestine G.E. 1988. Johan Stephan Rittangel's stay in the Netherlands (1641-2). In *Jewish-Christian relations in the seventeenth century: studies and documents*, ed. J. van den Berg and Ernestine van der Wall, 119–134. Dordrecht: Kluwer.

van Lieshout, H.H.M. 2001. *The making of Pierre Bayle's Dictionaire historique et critique*. Amsterdam: APA-Holland University Press.
van Lieshout, Leny. 2008. La vie culturelle à Rotterdam du temps de Bayle. In *Pierre Bayle (1647–1706), le philosophe de Rotterdam: philosophy, religion and reception*, ed. Wiep van Bunge and Hans Bots, 3–19. Leiden: Brill.
Vercruysse, Jerome. 1978. Davide, il re secondo il cuore di Dio, e l'illuminismo: Bayle, Voltaire, d'Holbach. In *La politica della ragione: studi sull'illuminismo francese*, ed. Paulo Casini, 233–248. Bologna: Mulino.
Whelan, Ruth. 2010. De Democritus et Haraclitus: Pierre Bayle et le rire. In *Les éclaircissements de Pierre Bayle*, ed. Hubert Bost and Antony McKenna. Paris: Champion.
Yardeni, Miriam. 1980. La vision des juifs et du judaïsme dans l'oeuvre de Pierre Bayle. In *Les juifs dans l'histoire de France*, ed. M. Yardeni, 86–95. Leiden: Brill.

Chapter 3
Philosophy and Its Islamic Moment: Pierre Bayle and Islam

Pierre-Olivier Léchot

Abstract Pierre Bayle paid particular attention to the Arabic scholarship of his century, as evidenced by the many articles in the Dictionary devoted to Orientalists. Yet his attention was not limited to a form of scholarly curiosity. On the contrary, Bayle made confrontation with Islam one of the keys to his reflection on religion as an anthropological phenomenon, to the extent that his commitment to the subject can be seen as a form of Islamic turn in Western philosophy. Islam thus constitutes a focal point for his reflections on tolerance, fanaticism and free thought, but also represents one of the paradigmatic places to which his historical method is applied, as far from reiterating all the legends about Muhammad, he intends to base himself primarily on sources rather than testimonies.

Keywords Islam · Anthropology · Tolerance · Fanaticism · Muhammad

If we consider the history of Orientalism, the last years of the seventeenth century constitute a real crucible—an "Islamic moment" to parody the title of a recent book (van Boxel et al. 2022). The year 1697 not only saw the publication of Bayle's *Dictionnaire historique et critique*, it was also the year in which another monument of erudition came off the press: the *Bibliothèque orientale ou dictionnaire universel* by Barthélemy d'Herbelot de Molinville. The following year, at the same time as the English theologian Humphrey Prideaux's influential biography of Muhammad was published in French by Daniel de Larroque, a close friend of Bayle, another

---

This article partly takes up some of the developments devoted to Pierre Bayle in Léchot 2021, 358–383. All Bayle's quotations are directly translated from French by myself. I would like to thank Mariepierre Bassac for her correction of my English.

P.-O. Léchot (✉)
Institut protestant de théologie, Paris, France
e-mail: pierre-olivier.lechot@ipt-edu.fr

© The Author(s), under exclusive license to Springer Nature Switzerland AG 2024
M. García-Alonso, J. C. Laursen (eds.), *The Importance of Non-Christian Religions in the Philosophy of Pierre Bayle*, International Archives of the History of Ideas Archives internationales d'histoire des idées 251, https://doi.org/10.1007/978-3-031-64865-6_3

monument was published in Rome that would influence the entire Enlightenment: the Arabic edition and Latin translation of the Qur'an by the Italian priest Ludovico Marracci. This short Muslim decade finally culminated in 1705 with the publication of the first edition of the *De Religione mohammedica libri duo* by the Dutch philologist Adriaan Reland, in which the scholar from Utrecht was one of the first to advocate the most objective possible approach to the Muslim religion.

It is therefore hardly surprising that Islam came to play a key role in the work and thought of Pierre Bayle. Although not himself an Arabist, he did not fail to take a close interest in the scholarly production on the subject of Islam and, more broadly, the Orient, as can be seen from the numerous articles devoted in his *Dictionary* to the Orientalists of his century as well as to a number of figures in Muslim and Arabic history. Among these, of course, Muhammad plays a central role, since the article devoted to him in the *Dictionary* is the second longest in the book, after that devoted to Spinoza (van der Lugt 2016, 35). In order to understand the role played by Islam in Pierre Bayle's philosophical thought, it is therefore crucial to situate it in the context of the development of Muslim and Arabic studies in the Grand Siècle.

3.1 The State of Knowledge About Islam in the Time of Bayle

Bayle was not the only philosopher of his time to be concerned with Islam: John Locke, for example, devoted several of his reflections on political theory or the question of tolerance to the subject (Matar 1991, 2015). What distinguishes Bayle's approach, however, is the extent of the erudition he displays in his works and, of course, particularly in the *Dictionary*.[1] Such a scholarly concern was not natural in the milieu of Reformed thinkers of his time. In many Protestant texts by writers of the Reformation and of the seventeenth century, many of the legends and anti-Muslim polemics developed in the Middle Ages were still alive.[2] Martin Luther did not hesitate to take up and even translate some of the most aggressive medieval treatises, such as *Refutatio Alcorani* by the Dominican Riccoldo da Monte Croce (1242–1320) or the *Treatise on the Manners of the Turks* by another Dominican whose apocalyptic views he shared: George of Hungary (c. 1422–1502). If, as we shall see, Protestant scholarship had begun to undermine many of the legends peddled by medieval polemicists, several of them were still very much in evidence among authors of the Grand Siècle. The most telling example maybe that of Gisbert Voetius (1589–1676), an orthodox Calvinist theologian and professor at the Academy of Utrecht (Léchot 2021, 227–229, 275–283). Although Voetius had a

---

[1] It is worth noting here a certain proximity between Bayle and Leibniz, whose interest in Islam was probably as great as Bayle's. See: Varani 2008.

[2] In the vast literature on the subject of the Western and Christian view of Islam, the main studies are: Daniel 1960; Tolan 2002; Burman 2007; Tolan 2019; Hanne 2019; Malcolm 2019; Léchot 2021.

very good knowledge of Arabic and was keen to study Muslim sources rather than rely on the opinions of authors who had studied Muhammad's religion, he nevertheless called on potential controversialists who were to debate with Muslims to demonstrate the superiority of the Bible over the Qur'an with arguments taken directly from Riccoldo's treatise, which Luther had already used.

The first literary genre to allow the development of a "new paradigm" (Malcolm 2019, 131–158) was travel literature. The number of travel reports in Islamic lands (Ottoman Empire, Persia, Syria, Egypt, Arabia) had indeed increased, to some extent as a consequence of the alliance between the France of François I and the Sublime Porte.[3] The first of these travellers, whom Bayle knew, was Guillaume Postel (1510–1581). He had accompanied the very first French ambassador to the Ottoman sultan, Jean de la Forest, and then travelled from Istanbul to Syria and Egypt after 1535. On his return to Paris, his knowledge of Arabic and his experience in the field earned him the reputation of being one of the best Arabists of his time— among his students was the future and famous Leiden philologist Joseph Juste Scaliger (1540–1609). However, it was not until 1560 that Postel published his first work on the Ottoman Empire.[4] Despite his scandalous reputation and his rather unorthodox views on many points of Christian doctrine, there is no doubt that Postel's writings contributed to a better understanding of life under the Sultans. In the wake of the eccentric Arabist, several travellers to the Levant wrote accounts of their journeys, including Jacques Gassot, Pierre Belon, André Thévet, and Nicolas de Nicolay. Their travel accounts thus gradually established a less polemical view of Islam, although these new views undeniably still coexisted with old prejudices. As Noel Malcolm points out, the reason for this different view lay in the personal experience that was at the origin of their stories (2019, 134). Of course, it would be wrong to claim that these narratives marked the beginning of a more objective look at conditions in the Muslim territories. But they did, however, mark the beginning of a paradigm shift. Increasingly, the Ottoman Empire was seen as a stable state, marked by order and discipline, and protecting several communities under the authority of the Turkish emperor. Of course, it was still argued that it was fear that enabled the Ottoman sultan to secure the obedience of his subjects. But later travellers, such as the Habsburg emperor's ambassador Ogier Ghislin de Busbecq (1522–1592), soon emphasised the importance of meritocracy within the empire (2010, 117–118 for example). The influence of these new narratives can be seen in the writings of political philosophers such as Jean Bodin (1529–1596) who, in line with many travel accounts, was led to highlight the tolerance practised by the Ottomans towards religious minorities within the empire (Malcolm 2019, 154). In the course of the seventeenth century, this travel literature evolved, as France began

---

[3] On travel to the Levant during the Renaissance period and the seventeenth century, see: Atkinson 1924; Cruysse 2002; Jaquin 2010; Richard 1995; Tinguely 2000. For an inventory of these travel accounts, see: Booromeo 2007.

[4] Postel 1560, 1575. See: Lestringant 1985; Balagna Coustou 1989; Le Borgne 2022; Tinguely 2022.

to send real archaeological missions to the Levant and even beyond.[5] Moreover, it does not focus exclusively on Islam but also includes many accounts of the diversity of Eastern Christianity—this is particularly true in the French Catholic world, which has close ties with Eastern Christians.[6]

The contours of Bayle's knowledge of travel literature are open to debate: on the one hand, Guy Stroumsa (2010, 126) has highlighted the importance of the account of Pietro della Valle (1586–1652)—a scholar, philologist, musicologist and composer who, from 1616 to 1628, undertook a journey that took him from Turkey and the Levant to Baghdad (where he married a young Baghdadi woman) and then to Persia and India. Other historians, on the other hand, have emphasised the decisive role of the account of the British envoy to the Sublime Porte, Paul Rycaut (Charnley 1998). This question can be answered if we focus on an article that makes particular use of travel literature: "Fatime". This article is, of course, about Fāṭima, the daughter of Muhammad and wife of ʿAlī ibn Abī Ṭālib. However, it is not her biography that is of interest to Bayle, but the fact that she has been identified by some travellers with the saint to whom the main mosque in Qom (Persia) is dedicated. Nevertheless, this identification is, as Bayle notes, a matter of contention among travel writers: Sir Thomas Herbert (1606–1682) identifies the saint with Fāṭima (DHC, Fatime, A), while Don García de Silva Figueroa (1550–1624) sees her as the daughter of ʿAlī and Fāṭima (DHC, Fatime, B) and Pietro della Valle and Jean-Baptiste Tavernier (1605–1689) consider her to be the granddaughter of the couple (DHC, Fatime, C). Bayle himself prefers to follow the Protestant Jean Chardin (1643–1713)[7] who reports a prayer pronounced in honour of the saint of Qom which identifies her with "the daughter of Mousa, son of Dgafer" (DHC, Fatime, D)—which is indeed the correct reading, the mosque being dedicated to Fāṭima bint Mūsā (790 AD–816 AD), daughter of the seventh Twelver Shīʿa Imam, Mūsā ibn Jaʿfar al-Kāẓim (745 AD–799 AD). This point underlines a well-known aspect of Bayle's historical method[8]: he prefers to trust a "document" (the prayer in this case) rather than the interpretation or testimony of travellers, whose opinion he disregards: "these are form prayers, and therefore they provide good evidence, which does not give us a great idea of the accuracy of the travellers, since some of the most famous ones report so poorly on the qualities of such a saint" (DHC, Fatime). However, the ultimate goal of the article is not the identification of the saint of Qom, but the highlighting of an observation of religious and anthropological nature to which the prayers celebrating her bear witness: as in Catholicism, Virgins go to paradise in

---

[5] Laurens 1978, 5. See for example the travel account of Hanna Dyâb (c. 1688-c. 1766), member of an important Maronite family from Aleppo and close to Antoine Galland, who accompanied the explorer commissioned by Louis XIV, Paul Lucas (1664–1737), during his journey and reported on his numerous archaeological acquisitions. See Dyâb 2015. On Lucas, see: van der Cruysse 2002, passim.

[6] See: Heyberger 2015; Girard 2011.

[7] On Chardin, see in particular: van der Cruysse 1998.

[8] On Bayle's historical method, see: Labrousse 1957, 1996, 3–68; Whelan 1989; Bost 1990.

Islam and certain saintly figures can be born of an "immaculate conception" (DHC, Fatime, D and E).

A careful study of the article's sources shows how Bayle works: although he quotes Tavernier and Chardin from their own accounts,[9] his knowledge of the other travellers' narratives cited in the article (including della Valle) comes from the "Remarques curieuses" written by Henri Bespier and attached to his translation of Paul Rycaut's *The Present State of the Ottoman Empire* (1677). A native of Sedan, Henri Bespier, sometimes spelled Bespierre or Bépierre (1627/8–1676),[10] had been the pastor of Senitot (near Harfleur, presently in the suburbs of Le Havre) since 1651. An expert in Oriental studies, he had been accused in 1673 of distributing a text favourable to the reunion of Protestants and Catholics. Published posthumously in 1677, his translation of Rycaut aimed to offer a more accurate version than the traduction of Briot (Rycaut 1677, "Avertissement") but also some 221 pages of his own composition as a commentary to Rycaut's text. In addition to various analyses, some of which are linguistic or cultural in nature, there are large extracts from travel literature, which makes it a small anthology of oriental travel accounts from which Bayle drew extensively and often. In fact, he had been aware of this work since his years in Sedan, insofar as he mentions it in a letter to his father in 1678.[11] It is therefore likely that most of his quotations from travellers to the Levant, both in the *Dictionary* and in his earlier works, are taken from them, to such a point that it answers Ann Thomson's question about Bespier's real influence (2017).

Of course, mention should also be made here of the work of Paul Rycaut (1629–1700), whose translation Bayle knew from Bespier (1677) and, most probably, from Briot (1670).[12] A Secretary to the English ambassador in Istanbul from 1661 to 1669 and then a consul in Smyrna, Rycaut published in 1668 a work that was to be one of the most influential presentations of Turkish manners and customs in the seventeenth century: *The Present State of the Ottoman Empire*. What makes Rycaut's book particularly important is that it was not a travelogue, but a religio-political analysis of the functioning of the Turkish state and culture based on his own experience in the land of the Sultan. His analysis was grounded on the conviction that the organisation of Ottoman society was built on specific political maxims: in a state as vast as the Sultan's empire, the ability to act quickly was a crucial point in the exercise of power (Rycaut 1668, 2). This explains, in Rycaut's view, the special position of the "Great Lord" whose political power can be considered "absolute" (Rycaut 1668, 6). For the Turks, the Sultan is indeed the equivalent of a god, the shadow of Allah on earth. In legislative matters, this power results in an association of the Sultan's word with law: "his mouth is the law it-self" (Rycaut 1668, 6). This does not mean, however, that the Sultan does not need to rely on religion for

---

[9] Tavernier 1676; Chardin 1686.
[10] Thomson 2017; Daireaux 2010, 615–616.
[11] Bayle to Jean Bayle, 26.11.1678 (*Correspondance* Nr 160) and also to Jean Rou, 21.11.1679 (*Correspondance* Nr 178). Bayle already refers to it several times in PD, as we shall see below.
[12] On Rycaut see: Anderson 1989.

political purposes, since he sometimes seeks the advice of the Mufti to secure his decision, even if this is a purely utilitarian point of view (Rycaut 1668, 6)—a Machiavellian perspective, therefore, that will not be missed by Bayle. Rycaut stands thus at the beginning of a long tradition of political reading of Turkish society that will be animated by readers of his work such as Montesquieu, Volney or Herder.[13] As we shall see again, Bayle was not insensitive to Rycaut's considerations and to some of the accounts reported in his work.

Alongside the travel accounts, a second field of study opened up at the end of the sixteenth century: that of Islamic and Arabic scholarship.[14] Sixteenth-century Europe's confrontation with the Ottoman Empire and Catholic polemics aimed at identifying the Reformation with Islam led Protestants to develop a specific interest for the religion of Muhammad. Despite the largely polemical and apocalyptic dimension of his views, Luther was not unimportant in the evolution of the study and dissemination of Muslim texts in the protestant European intellectual world.[15] In 1542, for example, he vigorously supported the publication of the Latin translation of the Qur'an by Robert of Ketton (twelfth century) to the Basel city authorities, which the Zurich humanist Theodor Bibliander (1505–1564) had begun to publish in the Rhineland city. Based on a sort of inversion of the Protestant principle of "Sola Scriptura", the scholarly study of Islam initially took the form of a study of it *as a text* (the Qur'an) or as a set of texts. While the first generation of Protestant scholars to take an interest in Islam (that of Scaliger, Bedwell, Raphelengius and Casaubon) concentrated mainly on a study of the Arabic language with a view to its mastery and use in the reading of texts (hence the establishment of handwritten Arabic dictionaries by most of these authors), the concern of the second generation was to make a vast literature accessible by means of editing and/or translating Arabic and/or Muslim sources. The Leiden Professor of oriental languages Thomas Erpenius (1584–1624) was the first to publish a sura from the Qur'an (the twelfth, "Joseph", *Yūsuf*; Erpenius 1617) as well as an Arabic-Latin *Historia saracenica* (1625) based on the coptic historian Jirjiis al-Makīn's (AD 1205-AD 1273) *Blessed Collection* (*Al-majmūʿ al-mubārak*), which Bayle used in his *Dictionary*, even devoting an article to its author (DHC, Elmacin).[16] Naturally, Catholic scholars contributed to this circulation of Muslim and more widely oriental sources in print, especially those of Eastern Catholic origin. For example, the Lebanese maronite scholar and professor at the Royal College in Paris, Gabriel Sionita (1577–1648), edited and translated under the title *Testamentum et pactiones initiae inter Mohamedem et Christianae fidei cultores* (1630) the text of an agreement of the Prophet with "the Christians of the world", the manuscript of which had been found in the monastery of Saint Catherine of Mount Carmel (English translation by

---

[13] See especially: Malcolm 2019, 289; Bevilacqua 2018, 173–175.

[14] For a good synthesis, see: Heyberger 2015.

[15] Bobzin 2008; Léchot 2021, 81–86.

[16] Erpenius 1625; on Erpenius, see: Vrolijk 2009; Vrolijk and van Leeuwen 2014, 31–40; Vrolijk and Weinberg 2020; Léchot 2021, 162–172 and 195–199.

Morrow 2015, 39–46). From the moment of its publication and for almost a century, the book provoked a wide-ranging discussion among scholars, some of whom could not admit its authenticity, others, on the contrary, seeking to prove it at all costs—a discussion which, if recent exchanges between Islamologists are to be believed, is still not over (Morrow 2016, 610–617). Of course, as we shall see, Bayle did not fail to address this matter.

In England, Arabic studies were mainly marked by the figure of Edward Pococke (1604–1691), the first professor of Arabic at Oxford University.[17] Pococke grounded his reputation on his editions of both Arabic and Jewish sources. Most of these publications were based on his rich collection of manuscripts, acquired while he was chaplain to the English Company of the Levant in Aleppo (1630–1636). In addition to his edition of Maimonides's commentary on the Mishnah (reconstituted on the basis of a series of manuscripts brought back from Aleppo, one of which was probably in the Jewish philosopher's own hand), Pococke became best known for his *Specimen historiae Arabum* (1651). This was a selection of extracts from the *Al-Mukhtaṣar fi-l-Duwal* of Abū al-Faraj (AD 1226-AD 1286), better known by his Latin name of Abulpharagius or Bar Hebraeus, one of the most erudite authors of the Syriac Orthodox Church. The specificity of Pococke's work lay in the abundance of his annotations, compared to the edited text: the original text and its Latin translation cover 31 pages while the annotations occupy nearly 350. In 1663 the entire book was to be published in Latin translation, again by Pococke. After al-Makīn, translated by Erpenius, Abū al-Faraj was thus the second Arab historian to appear in Latin in Europe and, logically, Bayle also devotes an article to him, which is indeed rather severe, since Bayle considers that he cannot be trusted when dealing with Greek and Roman history (DHC, Abulpharage).[18] Among the important Arabist figures of the century, Jacob Golius (1596–1667), "the man who had the most knowledge in the world of these [oriental] languages" (DHC, Hottinger), also played a key role through his editions but also, and above all, through the publication of his *Lexicon Arabico-Latinum* (1653), which Bayle praised (DHC, Golius, F).[19] A disciple of Erpenius, Golius, like Pococke, had passed through Aleppo, where he had also collected a number of manuscripts that would later enable him to establish his reputation as an Arabist. Golius was so influential in the scholarly world that Leiden became known as the "Mecca of Arabic studies" partly because of his teaching and the number of Arabists he trained there (Loop 2013, 15).

Among Golius's many students, the Swiss Johann Heinrich Hottinger (1620–1667) was also a central figure in Arabic studies, "one of the most famous writers of the 17th century", as Bayle calls him (DHC, Hottinger).[20] In contrast to

---

[17] Toomer 1996, 116–225; Gallien 2015; Mills 2020, 51–57, 71–95 and 205–218; Léchot 2021, 186–191; Williams 2022.

[18] See also his critic in DHC, Aurélien: "We should also count as a lie that Abulpharage says that Aurelian, in making peace with Sapor king of Persia, gave him his daughter in marriage."

[19] On Golius, see: Toomer 1996, 47–51; Vrolijk and van Leeuwen 2014, 41–48; Léchot 2021, 172–175.

[20] Loop 2013; Léchot 2021, 175–181, 204–206, 235–237 and 244–251.

Pococke or Golius, Hottinger distinguished himself less by his editions than by his synthetic studies (in fact mostly compilations) on the Orient. In addition to his *Historia Orientalis* (1651), which, as we shall see, was used by Bayle in a corrosive manner against Christianity, mention should also be made of his *Bibliotheca Orientalis* (1658), a vast bibliographical undertaking concerning documents, both manuscripts and printed, originating in the Levant, a significant part of which was contained in the Golius collection to which Hottinger had had access during his studies in Leiden. Despite his impressive erudition, Hottinger did not aim to develop a study of the Eastern world per se, but rather to show how knowledge about it could serve Protestant theology and apology. Hottinger was indeed a renowned polemicist as well as an apologist known for the breadth of his knowledge. But he is also representative of a trend that started at the beginning of the century and that was to find its illustration in the works of authors such as Samuel Bochart, Lodwijk de Dieu or in the biblical commentaries of the old Pococke. For them, Arabic could be used to understand the meaning of certain obscure biblical words or passages. For some of these authors, this use of Arabic was not without a certain form of anti-Judaism, or at least a criticism of the use of rabbinical literature by some Protestant Hebraists. This new trend in Protestant exegesis (which was again illustrated in the eighteenth century by authors such as Albert Schultens) was distinguished by a particular attention to etymology, to such an extent as to leave aside the concern for cultural recontextualisation shown by exegetes versed in rabbinical literature (Léchot 2023b).

At the end of the century, the study of Islam and, more broadly, of the Orient underwent a new evolution. After a century of publishing and "cataloguing" sources, the need for a critical synthesis of those gradually arose. The most important representative of this new trend was Richard Simon (1638–1712): through his *Additions to Brerewood* (published only in the twentieth century) and, above all, his *Histoire critique de la Creance et des Coutumes des nations du Levant* (1684), which took up the substance of the *Additions*, Simon intended to be less a scholar than a scholarly populariser[21]: as Jacques Le Brun pointed out, "the essential thing for him was not the 'invention' of documents, but the 'critique'" (Simon 1983, 35). Simon was one of the first to offer a picture of the religious situation in the Near and Middle East that was as complete and accurate as possible, without any form of value judgement on the religions treated, even if it were to be based on confessional reasons: "I do not know if one should give the name of piety to a thing that departs from the truth", he wrote already in the *Additions* (1983, 52). While the main part of the book focused on Eastern Christianity, it also included chapters on Islam and Eastern Judaism. In addition to a systematically critical look at the mostly polemical descriptions of Eastern Christianity and Islam given by several theologians, Simon's

---

[21] Also note the edition and translation made by Simon of Dandini's *Journey to Mount Lebanon* (Dandini 1675), which included many critical notes and which Bayle mentions in a letter to Vincent Minutoli: "The most singular thing about this writer is the observations he has made about the way in which the doctors of the Alcoran argue and dogmatise. There is, according to the Jesuit, a scholastic theology among the Mahometans, and an infinity of sects among the doctors." Bayle to Minutoli, 28.05.1675 (*Correspondance* Nr 93).

overview drew as much from Protestant as from Catholic scholarship. This critical view, more concerned with truth than with confessionnal dispute, did not escape Bayle in his review of the work in the *Nouvelles de la République des Lettres* of May 1684:

> He shows by convincing proofs, that the Latins have accused the Oriental Churches of an infinite number of heresies in a bad way and with much enthusiasm [...] The last chapter of this book deals with the religion of the Mahomedans and says some very particular and curious things about it without any concern (NRL: OD 1, 44 and 47).

In a certain sense, as we shall see, the first impulse that led Bayle to write his article "Mahomet" can be situated in the critical line inaugurated by Simon—and to which the very name of his *Dictionnaire historique et critique* testifies most clearly.

Before examining the Baylian view of Islam and its motivations, it is appropriate to look at three fundamental works that marked the last years of the seventeenth century and that I mentioned in my introduction to this paper. The first is of course the earliest translation of the Qur'an into Latin since the Middle Ages: that of the Italian priest Ludovico Marracci (1612–1700).[22] The need to provide the public with a new translation of the Qur'an had been on the minds of European scholars since the sixteenth century and the publication in Basel of Robert of Ketton's translation, which was already several centuries old. In 1647, a French translation of the Qur'an was published by André Du Ryer (1580–1660/1672), a diplomatic agent in Constantinople on behalf of the King of France and a later ambassador to Alexandria.[23] Banned by the censors under pressure from Vincent de Paul, Du Ryer's translation was nonetheless a great success, as it was reprinted many times and was itself translated into several languages. Although Bayle devotes an article to him in the Dictionary, this article is relatively brief and emphasises that several scholars have criticised his translation methods (DHC, Ryer (André du), B). Of course, this French translation could not satisfy a literate public, anxious to have both an accurate Arabic edition (which was far from obvious because of the scarcity of complete manuscripts circulating in Europe) and a Latin translation as close to the text as possible. Numerous attempts were made throughout the century, both on Protestant and Catholic sides, with one group competing with the other in a race to translate the Qur'an. Probably because he had the benefit of a vast series of manuscript sources (in particular Qur'anic commentaries or *tafsīr*), Marracci won this marathon with an edition/translation that would prove decisive throughout the eighteenth century. In 1691, he had already published a *Prodromus* to his translation which Bayle seems to have known only through a review published in the *Acta Eruditorum* of 1692.[24] In fact, nothing seems to confirm that Bayle ever consulted Marracci's translation, as his references to the Qur'an in "Mahomet" in the 1702 edition of the *Dictionnaire* are mostly based on Bibliander's Latin edition (1543) or on Du Ryer's translation.

---

[22] Rizzi 2007; Glei and Tottoli 2009; Bevilacqua 2013, Bevilacqua 2018, *44–74*.
[23] Hamilton and Richard 2004; Larzul 2009; Vigliano and Wele 2021.
[24] DHC, Mahomet.

As with Marracci's translation, the first edition of the *Dictionnaire* (1697) could not take into account the two other major publications of the end of the century in the field of Oriental studies: the biography of Muhammad by the Anglican priest Humphrey Prideaux and the *Bibliothèque orientale* by the French orientalist Barthélemy d'Herbelot. Coming from a family of the Robe, Barthélemy d'Herbelot de Molinville (1625–1696)[25] had studied oriental languages before becoming a close friend of Nicolas Fouquet and of the Grand Duke of Tuscany with whom he stayed (1655, 1666) before returning to France with oriental manuscripts which were of great importance in the development of his magnum opus: The *Bibliothèque orientale*, which was published after his death thanks to the involvement of another outstanding Arabist, Antoine Galland (1646–1715), who would be the translator of the *Thousand and One Nights* in French. Since its publication, the work has been the subject of numerous criticisms, including that of Edward Said, who reproached it for having in a way corseted oriental knowledge by transforming it into "a rational panorama of the Orient, from A to Z" (Said 2003, 129). It has since been shown that this "encasement" was not the fruit of the Orientalist, but that of his main source: the bibliographic encyclopaedia of the Ottoman scholar Kâtip Çelebi or Ḥājjī Khalīfa (1017 AH/1609 AD-1068 AH/1657 AD).[26] Of course, d'Herbelot also had recourse to other sources and, in particular, several anthologies of Persian origin.[27] While invalidating the Saidian thesis, this remark also allows us to underline that the work did not subscribe to an orientation favoured by Bayle (following Simon), namely the properly critical dimension of their work (Dew 2004, 252). D'Herbelot in fact belongs more to the group of scholars of the first part of the century like Hottinger, his almost contemporary, whose objective was more to compile sources according to the practice of antiquarians than to propose a critical reading of the latter through their comparison and evaluation.[28] The book is distinguished, however, by the central role given to Islam, since it constitutes the common point between the different cultures examined: Turkish, Persian and Arabic. Jean-Baptiste Dubos was not mistaken when he described the work as a "Mahometan library" in several of his letters to Bayle.[29] This does not mean, however, that d'Herbelot was showing any form of Islamophilia—on the contrary, his work, like Bayle's,

---

[25] See in particular: Laurens 1978; Dew 2004, 2009; Bevilacqua 2016.

[26] Dew 2004, 238–240; Laurens 1978, 50.

[27] Laurens 1978, 59. See also Torabi 1992.

[28] While it is not wrong to note a certain closeness between the Bibliothèque and Bayle's Dictionnaire (Dew 2004, 247), it is also necessary to highlight the differences between the two (Dew 2004, 249). One of the problems posed by any dictionary is indeed that of a potential discontinuity between alphabetically ordered articles. This is the case with d'Herbelot, which Bayle does not fail to note by pointing out the latter's "incomplete narrative", even if it is to immediately nuance his judgement: "this advice is important to all the authors of a Dictionaire, and it is very difficult for them not to fall into this fault. I fear that it has escaped me more than once." (DHC, Mahomet, PP, pour un exemple).

[29] Dubos à Bayle, 19.12.1695 and 10.08.1696 (*Correspondance*, Nr 1067 and 1148).

emphasised the impostor character of Muhammad. Despite these differences, Bayle did not fail to refer to the *Bibliothèque* in the second edition of his *Dictionary* (1702).[30]

The work of Humphrey Prideaux (1648–1724), a former pupil of Pococke and Dean of Norwich, was also to be a landmark, especially in the eyes of those close to Bayle who, like Matthieu Marais, saw in the article "Mahomet" of the *Dictionnaire* a "good commentary on this book, which appears to be quite short in places and too long in others."[31] Written primarily for the purpose of contemporary ecclesiastical polemic (especially against the deists), the work was not lacking in scholarly foundation and indeed included in its conclusion an annotated bibliography of the principal sources enabling the reconstruction of the Prophet's life (Mandelbrote 2022, 262). The *True Nature of Imposture* nonetheless took up many of the criticisms levelled against Islam since the Middle Ages and stressed, among other things, that the poetic quality of the Qur'an implied that it could not have been written by an illiterate person, as the Prophet of Islam was described. On the contrary, the holy book was the product of (heretical) Christian and Jewish scholars employed by Muhammad, and the emergence of Islam from a corrupted Eastern Christianity was to serve as a warning to all those in England who might consider being unfaithful to the tradition of their Church. As Scott Mandelbrote notes, Prideaux's work was not so much symptomatic of an orientalism interested in the East or Islam as such, but rather of an Arabist scholarship eager to show its importance in the contemporary Christian polemical and apologetic context (2022, 263). From a certain point of view, the same could be said of Bayle's enterprise—still bearing in mind all that separates it from that of the English clergyman. The fact remains that Bayle found a certain interest in it, since he advocated in favour of its translation by one of his correspondents: "I have been assured that there are some very curious things in this Life of Mahomet and I wish it to be translated into Latin, or into French."[32]

3.2 From Polemical Legends to the Anthropological Reading of Religions

Bayle stands at the confluence of these different currents: travel narratives, Arabic scholarship and critical reading—and, logically, he draws the consequences on the historical and philosophical level. As the title of the *Dictionnaire* indicates, it is

---

[30] See for example the article Abudhaher.

[31] Mathieu Marais to Bayle, 21.02.1699 (*Correspondance* Nr 1414): "Your article on 'Mahomet' is a good commentary on this book, which seems very short in places and too long in others. It was not necessary otherwise that Dr. Prideaux compared the Alcoran to the works of Homer to teach us that the Alcoran was composed of pieces collected as the poem of Homer was formerly done on the centos and rhapsodies which ran under the name of this poet."

[32] Bayle to Dubos, 24.06.1697 (*Correspondance* Nr 1270) An account of the English version appeared in Basnage de Beauval's *Histoire des Ouvrage Savants* (October 1697, art. VII).

certainly a historical undertaking, but it also intends to use the tools of *criticism* as defined by Simon to re-establish the facts, including when it comes to revoking certain legends forged by Christians for the sake of their theological and confessional prejudices. Bayle thus intends to dissociate the current legends about the Prophet from what we can learn about him by consulting his writings. For him, the contradiction between the works and sayings of Muhammad on the one hand and the legends propagated about him on the other must always be resolved in favour of the former. This is a question of epistemological ethics, especially with regard to historical actors:

> I would not wish to deny that in certain respects the zeal of our disputants is unjust; for if they make use of the extravagances of a Mahometan legend to render Mahomet himself odious or ridiculous, they are violating the fairness which is owed to everyone, to the most wicked as well as to the good people. One must never impute to people what they have not done; and consequently it is not permitted to argue against Mahomet by virtue of the reveries that his followers attribute to him, if it is not true that he himself has uttered them (DHC, Mahomet, H).

This is what leads Bayle to reject legends such as those, for example, of the pigeon speaking in the Prophet's ear, of his corpse being eaten by dogs, or of his tomb being suspended in the air by means of a magnetic mechanism intended to deceive overly credulous followers. With regard to this last legend, Bayle notes that it is a story that can be found in connection with almost all religions, but which is hardly believable "because of the considerable distance [...] between the iron statues and the stones that attracted them" (DHC, Mahomet, EE and FF).[33]

Behind the critical, factual and historical rigour, the anthropological analysis of religion and the concomitant criticism of Christianity that Bayle undertakes are therefore already apparent. Unlike Jesus, Muhammad did not perform any miracles and never claimed to do so, even to the point that he "often confesses in the Qur'an that he could not perform miracles" (DHC, Mahomet, H).[34] This did not prevent his followers from attributing to him certain miracles whose falsity has been exposed by generations of Christian polemicists. But this, for Bayle, is a tendency that is not peculiar to Islam, since similar phenomena can also be found, for example, in Catholicism. However, one should not conclude from this a general criticism of Islam as well as Catholicism, since these legends are the product of "nothing" authors and the "serious" writers do not mention them. While questioning a theological and confessional reading of the facts, Bayle also draws a systematic conclusion from his remarks: religion always sees the coexistence of a reading that one

---

[33] See DHC, Mahomet, H (for the legend of the dove) and Z (for the story of Muhammad's corpse devoured by dogs).

[34] Bayle takes up an old argument of the medieval criticism towards Islam. In this sense, he does not follow Vanini, who, like Naudé, had rather taken up the thesis of "false miracles" set up by Muhammad in order to convince the masses. Vanini's aim was to highlight the falsity of all miracles, Muslim or Christian. See: Khayati 2009, 120–121. On Naudé, see: Schino 2020, 104–105. If Bayle, like Vanini and Naudé, seeks to challenge Christian exceptionalism, he does not do so along the same lines. In this sense, we must nuance the statement of Khayati 2009, 122, which places Bayle too precisely (in my view) in the line of Vanini.

would readily describe as popular, inclined to seek out or even invent the miraculous, and a more intellectualist or even philosophical approach that does not stop at the supernatural (DHC, Mahomet, H and I).[35]

All these developments are thus set against the background of a confrontation of positive religions, their systems of thought and their hold on the reality of human societies. This is confirmed when Bayle addresses the question of the causes for the spread of Islam. In his view, Islam is indeed the most widespread religion on earth (DHC, Mahomet, A), which enables him to deal a fatal blow to the argument of Catholic universalism, which was then regularly put forward by Roman polemicists against the representatives of the Protestant "minority". Such an argument, apart from being contradicted by the facts (DHC, Mahomet, P),[36] is in reality rather questionable from a moral point of view: "What a great advantage to be able to understand the art of killing, bombing, and exterminating mankind much better than they do!" (DHC, Mahomet, P). For Bayle, there is no doubt that Islam owed its success to military force—just like Catholicism. This does not prevent him from examining other hypotheses which, of course, must be discarded, but which also provide the opportunity for new comparisons between Islam and Christianity.

First of all, Bayle intends to show that the argument according to which Islam has spread because of the attraction of its loose morals is completely unfounded.[37] This argument does not stand up to the facts, simply because the Qur'anic morality is anything but licentious for those who examine it carefully:

> You will find in it all that is most opposed to corruption of the heart; the precept of patience in adversity, that of not slandering your neighbour, that of being charitable, that of renouncing vanity, that of not doing wrong to anyone, and finally the one that is the abridgment of the law and the prophets, 'Do unto your neighbour as we would have you do unto us'. It is therefore an illusion to claim that the law of Mahomet was established with such promptitude and extent only because it took away from man the yoke of good works and painful observances, and allowed him bad morals. (DHC, Mahomet, L).[38]

As for the practice of this morality, it is no less followed by Muslims than that of Christ is among Christians: "I do not claim that Christians are more erratic in their morals than Infidels; but I do not dare affirm that they are less so" (DHC, Mahomet, P). In a reflection that was to have a bright future and that probably came from Jean Bodin[39] or Chardin,[40] Bayle only considers the influence of the "climate" prevailing in certain regions of the globe to explain certain variations in the application of the

---

[35] On this point, see later in this chapter.

[36] See RQP: OD 3, 706, where Bayle notes that in that case, it is indeed the doctrine of the Socinians according to which Christ is not God that has been most widely disseminated, since it is the one adopted by the Muslims. See also DHC, Mahomet II, D.

[37] The argument was an old one, since it can already be found in the writings of Pope Pius II. See Malcolm 2019, 34.

[38] See also DHC, Mahomet, II.

[39] On Bodin, see: Malcolm 2019, 217–220. On Herder's theory of climate, for example, see: Léchot 2021, 471–498.

[40] See in particular: Van der Cruysse 1998, 26 and 58.

morals of one or other of the two religions. It must be acknowledged that Muhammad preached a rigorous morality and was followed in this, despite his own erratic morals—proof, according to Bayle, that

> when one is once forewarned of the opinion that a certain man is a Prophet, or a great Servant of God, one believes rather that crimes are not crimes when he commits them, than one persuades oneself that he does a crime. This is the effect of the foolish prejudice of many small minds. [...] Every day we see diminutions of this prejudice: if a man has once acquired the reputation of a great zealot of Orthodoxy, if he has made his mark in battles against heresy offensively and defensively, you will find that more than half the world is so prejudiced in his favour that you cannot make them admit that he is wrong, by doing things which they would condemn if another did them. (DHC, Mahomet, II).

Is the reason for Muhammad's success to be found in the promise of a paradise of carnal pleasures? Again, this is an answer that Bayle rejects. For it is not the origin of pleasure (the senses rather than the mind) that causes men to adhere to one image of paradise rather than another, but the degree and quality of the promised pleasure (DHC, Mahomet, M). In this sense, the Christian paradise, though more spiritual, is more successful with the masses than that of Muhammad.

The explanation of Islam's successes through its use of armed force is therefore the most plausible, but it does not explain everything, starting with the reason why it is so difficult to convert Muslims while many Christians willingly and freely become Muslims (DHC, Mahomet, DD). On this point, Bayle disagrees with the distinction proposed by Riccoldo da Monte Croce, who asserted that the Muslim will become a Christian when death approaches, while the Christian will more willingly become a Muslim during his earthly life. To give credence to such a hypothesis would in fact be tantamount to recognising "that the Mahometan religion is more convenient for living, and that the Christian is safer for dying", which would entail a certain relativity of Christianity. Bayle's explanation is no less critical: in reality, "everyone wishes to die in the Religion in which he has been brought up": "Ignorance does [therefore] in the hearts of these Infidels what science produces in the heart of an honest Orthodox man, I mean an invincible attachment to his opinions." Attachment to religion is therefore *in fine* a matter of prejudice of birth and custom (DHC, Mahomet, DD).

The importance of prejudices in the attachment to a religion can be seen when it comes to the confrontation of its sacred texts with those of others: in this case, the Bible and the Qur'an. In Bayle's time, many Protestant theologians still believed that it was possible to win the battle of the Scriptures.[41] Gisbert Voetius, but also authors as different as Johannes Hoornbeeck (1617–1666), professor in Leiden, or the famous pedagogue Jan Amos Comenius (1592–1670) were all convinced that it was possible to demonstrate the superiority of the biblical text over the Qur'anic one: better doctrines, sounder moral principles, and more realistic hopes were only some of the many qualities that the biblical text could put forward in its confrontation with the Qur'anic revelation. Naturally, there was some naivety in thinking that one could convert Muslims to the thesis of unaltered biblical truth when Muslim

---

[41] See Léchot 2021, 260–288.

polemical theology had for centuries been repeating the thesis of the corruption of Jewish and Christian Scripture (*taḥrīf*). We can also measure here the permanence, among Protestant theologians of the time, of the medieval argument and, in particular, that of Riccoldo da Monte Croce (Léchot 2021, 281). It is striking to see that, on this point, Bayle's philosophy of religion preserved him from the illusion of a possible demonstration of the superiority of the biblical narrative. On this particular point, he relied in the article "Mahomet" on the arguments of a Muslim polemicist of which he had found a summary in Hottinger's *Historia orientalis* (Léchot 2022, 151–152): a certain Ahmed ibn-Edris, better known by his Arabic name of Shihāb al-Dīn al-Qarāfī (died AD 1285).[42] A famous Melkite jurist of Berber origin who settled in Egypt, al-Qarāfī was known in Islamic lands as one of the greatest anti-Christian polemicists. On the basis of what he could read in Hottinger's writings, Bayle stressed that Muslim controversialists had been pointing out the inconsistencies of Scripture for a long time and that these arguments were no less admissible than those of Christians against the Qur'an:

> There are Arabs who have written in favour of the Alcoran, and against the Bible, with enough industry to foment prejudices. Hottinger speaks of an Author who peels away the apparent contradictions of Scripture, and who even claims to prove by the Bible, the mission of Muhammad. We would be very simple, if we believed that a Turk, who examines this, finds it as weak as we do. He sees no force in the objections against the Alcoran; he sees much in the objections against the Christians. So great is the force of prejudices. (DHC, Mahomet, DD).

Again, Bayle emphasised that what guided men were not their principles (in this case, those of argumentative reason) but the prejudices that their religious education and partisan spirit had strongly rooted in their minds.

3.3 Muhammad: Impostor or Fanatic? Jurieu's Shadow

Beyond these anthropological and corrosive reflexions, however, the image of the Prophet that emerges from the reading of the article that Bayle devotes to him seems largely negative. Following Humphrey Prideaux, Bayle is of the opinion that Muhammad was an impostor, a "seducer" (DHC, Mahomet, R) and a "false prophet" (DHC, Mahomet, T and V) who would not have invented a revelation, forged from "several pieces of Judaism and Christianity" (DHC, Mahomet, N), had he known that he would have at his disposal troops entirely devoted to his project of dominating Arabia. On this occasion, Bayle proposes a rule to distinguish the impostor from the true prophet:

> A good touchstone for knowing whether those who boast of Inspirations, either for the purpose of spouting new Prophecies, or of explaining old ones, such as Revelation, are proceeding in good faith, is to examine whether their doctrine changes course in proportion as the times change, and that their own interest is not the same as before [...] One of these

---

[42] Van Koningsveld 1996; Champion 2010.

evidences is that the variations of his Prophetic Spirit responded to the change of his particular interests. (DHC, Mahomet, T).

The wording of this rule shows that Bayle's horizon was rather limited, at least at first sight: his reflection on Muhammad's "fanaticism" or "imposture" is indeed situated in the wake of his controversy with Pierre Jurieu (van der Lugt 2016, 131). The image of Muhammadan psychology proposed here corresponds roughly to Bayle's image of Jurieu, namely that of a religious man who appears orthodox but who can change his doctrine according to the evolution of events. In the past, Bayle had already taken pleasure in underlining Jurieu's changes of opinion, precisely on the subject of the Apocalypse: his former colleague had indeed predicted, on the basis of his reading of the last book of the Bible, the re-establishment of Protestantism in France for 1689, before retracting his prediction once it had been disavowed by events. It is also in the light of his controversy with Jurieu that we must understand his remark concerning the distinction to be made between those who "predict by credulity and illusion", those who "predict by politics" and finally the "Prophets of sedition", Jurieu probably being classified in the third category while Muhammad would be more appropriately classified in the second.

Indeed, in the GG remark of "Mahomet", Bayle lists all the erroneous predictions about the end of the Ottoman Empire or the conversion of Muslims common among Christians and Muslims alike, as well as those announcing the triumph of Islam. Bayle's reading of these prophecies is anthropological in nature: they can be explained, in part, by "the desire to console oneself, by the hope of the ruin of a furious persecutor, [which] makes it easy to find this ruin in the predictions of Scripture, or in some other sources." This is the prediction "by *illusion*" (emphasis added). There may also be *political* or *military* reasons, in which case the prediction is intended to galvanise the troops in their fight against the Turks. Finally, the prediction may also be intended to provoke an uprising. In this latter case, it is appropriate to speak of "prophets of *sedition*" (emphasis added). Bayle's conclusion is clear: "those who meddle in revealing the future to us in relation to the Turk, use their time badly" (DHC, Mahomet, GG).

If Bayle seems to hesitate between fanaticism and imposture concerning Jurieu, the same cannot be said of his perception of Muhammad, even if doubts appear here and there, and one may wonder whether they are not the result of his questions about his immediate opponent.[43] For Bayle, it is clear that Muhammad invented certain rules concerning polygamy or the right to marry solely in order to "colour his incontinence" with some religious virtue" (DHC, Mahomet, T). The author of the *Dictionnaire* thus devotes a long and detailed description to the sexuality of Muhammad, a man of "natural vigour, which was very surprising" (DHC, Mahomet, D) and who "performed the conjugal function with great strength" (DHC, Mahomet,

---

[43] See DHC, Mahomet, N, where the conditional formulation should be noted: "If anything *would* make me believe that there was fanaticism in his act, it would be to see an infinite number of things in the Alcoran that could only seem necessary if one did not want to use constraint. Now there are many things in this work which have been done since the first successes of Muhammad's arms."

S). While the Prophet never used his influence over women to propagate his religion, he did not hesitate to use them "for natural use, for the remedy of his incontinence, for venereal pleasure, in a word" (DHC, Mahomet, E).[44] These lengthy developments clearly indicate the importance, in Bayle's eyes, of the sexual phenomenon when it comes to historical causality (van der Lugt 2016, 47), but their main function is to support the thesis of the imposture of Muhammad:

> he used religion only as an expedient to enlarge himself. [...] Did a true fanatic ever have such a character? Does he understand his world so well? Would not a man who believed for some time that God sent him his Angel to reveal to him the true religion, be disillusioned when he found that he could not justify his mission by any miracle? [...] This does not smell of fanaticism. (DHC, Mahomet, K and T).

Associating Muhammad with a mere impostor was not, however, such an obvious reading, especially in view of the theological burden that had been placed on this figure by a whole Protestant tradition to which we have already alluded. To call the Prophet an impostor was to renounce seeing him as a dangerous fanatic, an instrument of the devil or even a figure of the antichrist, even though the medieval tradition had largely insisted on this point. Already in his *Janua coelorum reserata* (1692), Bayle had refused to see in Muhammad the Antichrist, for mainly hermeneutical reasons (JCR: OD 2, 872). In the *Dictionary*, Bayle again refuted the idea that Muhammad was the antichrist and refused to believe that the Reformers could have thought so (DHC, Mahomet, Y).

3.4 Tolerance and Violence: Islam Better than Christianity?

All the elements of the Baylean approach to Islam that we have just highlighted underline the extent to which religious history and philosophy of religion are intimately intertwined in his work and thought. Historical criticism, the establishment of facts, is the consequence of a philosophical ethos that places "equity" and "honesty" at its heart. But at the same time, historical criticism is also the condition for a philosophy of religion that is grounded on facts and not on mere illusions. It is therefore not surprising that Bayle's reflection on what he considered to be a crucial issue—that of religious tolerance and violence—is to be found precisely in the pages he devotes to Islam—and vice versa! In the *Critique générale* (1682) for example, Bayle was already attacking the violence that Catholic states have used over time precisely in relation to Islam:

---

[44] See also remarks Q and OO, which deal with the fate of women in Islam, as well as the one (PP) that Bayle devotes to the role of Aïsha in the development of the schism of ʿAlī. The schism was based entirely, in the eyes of Bayle, who follows Prideaux here, on the hatred of the Mother of Believers for the Prophet's son-in-law, who had denounced her "gallantries" with other men to her husband. Bayle also believes that it was his wives who murdered Muhammad, because of his infidelities and his propensity to beat them (DHC, Mahomet, E)—a thesis that is not so far from certain contemporary hypotheses; see: Ouardi 2017, 154–178.

> It is true that the Pope's religion governs itself with respect to all other Christian societies just as the Turks have governed themselves with respect to all the kingdoms and republics that have been within their reach. They did not grace any of them, they had to put everything under the yoke of their barbaric domination. In the same way the Catholic religion par excellence suffers no other, it overwhelms them all and massacres them all when it can, and, what is far more tyrannical, it gives no other reason for what it does, except that it is infallible. (CG: OD 1,77).

It is therefore all the more unfavourable to Catholicism to note that, in fact, on questions concerning individual conscience, the Turks behave in a much less intolerant manner:

> However odious this parallelism may be, it is nevertheless true that the Turks, Turks though they are, worry people less about matters of conscience than does the Roman Church. (CG: OD 1,77).

This double allusion to Islam as a religion that is both violent (in terms of its expansion) and tolerant (when it comes to the management of religious difference in the territories under its control) allows Bayle to underline, as by a mirror effect, the doubly scandalous character of Catholic policy: not only is the latter just as odious as those of the 'Turks' (think, says Bayle, of the crusades), but it is even more so since the Muslim empires tolerate the presence of other religions in their midst, once their authority is well established—something that Catholic states do not do with regard to Christian dissidents, driven as they are by their need to dominate not only bodies but also consciences. It is therefore not surprising that the St. Bartholomew's Day massacre alone (1572) caused more victims than all the Muslim persecutions against Christians and Jews combined (DHC, Mahomet, AA).

Of course, Bayle was not so naïve as not to question the motives of this Islamic tolerance and its relation to the doctrine contained "in the Alcoran". This is the reason why he carefully examines the authenticity of Muhammad's pact with the Christians of the world, edited by Gabriel Sionita in 1630. His conclusion is meaningful: if the pact is indeed authentic, it is nonetheless based on strategic motivations. At a time when Islam was still "weak and in its infancy" (militarily and politically speaking), it was preferable to favour religious tolerance rather than persecute other religions (DHC, Mahomet, AA and BB). Here Bayle resorts to a principle of Machiavellian inspiration stated in the *Pensées diverses*:

> It is to want to join together two incompatible things to want to be a conqueror and a persecutor of other religions because the peoples one wants to subjugate resist like lions when they know that one wants to force them into cults they believe to be evil. (PD: OD 3, 147).

Beyond this political interpretation, what matters to Bayle is that Muhammad did practice a form of tolerance towards Christians that formally contradicts the teachings of the Qur'an on this subject—Bayle refers here, according to a long Christian polemical tradition, to Sura 9.[45] For those who look at "matters of facts" (the only way to really judge the value of a religion in Bayle's eyes), Islam cannot be

---

[45] DHC, Mahomet, AA, and PD: OD 3, 147. For a summary of the Christian polemical use and abandonment of this reading of Sura 9, see Léchot 2023a.

associated with a purely violent religion. However, it is clear that the same cannot be said of the Christian religion—worse: Christians "have only been ordered to preach and instruct, and yet from time immemorial have exterminated with iron and fire those who are not of their religion" (DHC, Mahomet, AA). Behind the highlighting of the Islamic religion, a rather virulent criticism of Christianity and its missionary undertakings emerges:

> One can be very sure that if the Christians of the West had dominated Asia, instead of the Saracens and Turks, there would be no trace of the Greek Church there today, and that they would not have tolerated Mahometanism as these infidels have tolerated Christianity there. (DHC, Mahomet, AA).

Hence this classic conclusion from Bayle's pen, to the point of constituting an axiom of his thought: "men conduct themselves little according to their principles" (DHC, Mahomet, AA).

It would be wrong, however, to think that Bayle addressed such criticisms only to Catholics, insofar as it was a question of turning the defenders of coercion, Catholics and Protestants alike, back to back. In his line of fire are naturally authors such as Pierre Jurieu and his defence of the right of sovereigns to compel consciences (van der Lugt 2016, 36). From a historical point of view, it is clear that Islam spread through arms, whereas the Christian religion of the first three centuries did not use either political power or military force (DHC, Mahomet, O)[46]—it was only when it became the official religion of the Roman Empire that Christianity was able to establish itself with the help of the Roman emperors and later the Frankish kings.[47] This observation is naturally to the advantage of Christianity but the question remains: is this enough to support the thesis of the superiority of Christianity over Islam? Not at all for Bayle: if coercion could be judged positive from the first century until his time, it is because its legitimacy was based on a word of Christ himself: "compel them to enter" (Luke 14:23). Indeed, this is what Jurieu, like Bellarmine, recognises by giving as examples of legitimate coercion those of the kings of Israel and the Christian emperors, but also, precisely, certain Protestant princes. It is thus clear that each of these authors "establishes as an immutable principle of all times that the way of authority is right for the propagation of the faith" (DHC, Mahomet, O). Imagining the reaction of a Muslim to such an reasoning, Bayle had him address these words to Jurieu: "You should therefore, if you could, have used constraint from the day after the Ascension" (DHC, Mahomet, O). Jurieu may well be right in his demonstration, but he should nonetheless, "if he entered into a dispute with the Mahometans", renounce "the arguments that have always been provided against them by the way their religion has spread" (DHC, Mahomet, O). Of course, such a tactical choice would be a big mistake: coercion, of whatever religion, is wrong, even if Jurieu claims to defend it on the basis of biblical revelation and the Holy Spirit. History shows that violence is the best way to propagate a religion, but this can only be done at the price of sincere conversion:

---

[46] See CP: OD 2, 387.

[47] Here, Bayle refers to Jurieu 1687, esp. 280 and 289.

> Ask the French dragoons who served in this profession in 1685: they will tell you that they are determined to make the whole world sign the Alcoran, provided that they are given time to apply the maxim: compelle intrare, force them to enter. (DHC, Mahomet, N)[48]

In a prophetic passage, Bayle envisages the rather dark consequences of the logic of constraint put forward by Jurieu: "[Christians] will make a beautiful manège in the Indies and in China, if ever the secular arm favours them there. Be sure that they will use the maxims of Monsieur Jurieu there. They have already done so in some places" (DHC, Mahomet, AA).

The unique alternative to the logic of constraint can therefore only be tolerance, as Bayle points out as early as in the *Commentaire philosophique* (1686). Bayle will no doubt be reproached for his thoughts about tolerance "that not only Socinians, but also Jews and Turks should be suffered in the Republic" (CP: OD 2, 419). Such a conclusion is indeed well-founded, he notes, but it is not in itself absurd: "in this encounter, no middle ground can be found; it is all or nothing." With an uncommon radicality that contrasts, for example, with Locke's reservations, Bayle thus asserts that one "cannot have good reasons for tolerating one sect, if they are not good for tolerating another." Tolerating all Christian sects within a territory necessarily leads to accepting the presence of other religions such as Judaism but also Islam. Moreover, Bayle notes, Muslims are "more worthy of tolerance" than the Jews, since they recognise Christ as a great prophet. But it is, in fact, mainly a principle of reciprocity:

> If the Mufti should take the fancy of sending to Christendom some Missionaries, as the Pope sends to the Indies, and if these Turkish Missionaries were found insinuating themselves into the houses, in order to act as converters, I do not think that they would be entitled to be punished; for if they answered the same things that Christian missionaries would answer in Japan in such a case, namely that the zeal to make the true religion known to those who are ignorant of it, and to work for the salvation of their neighbour, whose blindness they deplore, has led them to come and share their light with them, and that without taking this answer into consideration, or hearing their reasons, they would be hanged, would it not be ridiculous to find it wrong for the Japanese to do the same? (CP: OD 2, 420).

This is a biblical teaching: "We must not forget the prohibition of double weights and measures, nor that with the same measure we measure others, we shall be measured" (CP: OD 2, 420).[49] And let it not be imagined that such tolerance on the part of Muslim missionaries would be to the detriment of Christianity! For the confrontation would inevitably turn to the advantage of the Christian evangelists: "Pagan and Mahometan preachers would gain nothing among us, and ours could do much fruit among the Infidel nations" (CP: OD 2, 420). The comment, of course, does not lack irony, since it points to the weak conviction of those who do not believe that the Christian religion "is capable of doing anything on its own" and believe that it is appropriate to give it "for Assistants the Executioners and the Dragoons, Assistants who do well without the truth, since on their own and without it they do what they want." Religion, in fact, is "a matter of conscience which cannot be

---

[48] See also: CP: OD 2, 400.
[49] It is a reference to Matthew 7:1–2 and Mark 4:24.

ordered"—tolerance is therefore due to Muslims as it is to all Christians, all the more so since the followers of Islam "have preserved for the Christians of their Empire the faculty of exercising their Religion" (CP: OD 2, 420). It is not for us to explain here how such a statement about religion as grounded in conscience can coexist with the one we noted above about the close links between religious conviction and prejudices of birth or custom. What is certain, however, is that the Baylian approach to Islam is by no means a simple synthesis of the achievements of Arabist scholarship of the Grand Siècle. The best proof of this is the way in which Bayle integrates Islam into his reconstruction of the history of free thought.

3.5 An Islamic Genealogy of Free-Thinking from Averroes to Spinoza

As early as 1682, Pierre Bayle was interested in the torment of Mahomet Efendi reported by Paul Rycaut in his book on the Ottoman Empire (Rycaut 1668, 130). In a significant development aimed at showing that atheism was not incompatible with the idea of honesty, he cited the example of the atheist philosopher Lucilio Vanini (1585–1619), who died at the stake in Toulouse, and continued:

> To the example of Vanini may be added that of a certain Mahomet Efendi, who was executed at Constantinople not long ago for having dogmatized against the existence of God. He could save his life by confessing his error, and promising to renounce it in the future: but he preferred to persist in his blasphemies, saying, that though he had no reward to expect, the love of truth obliged him to suffer martyrdom in order to uphold it. A man who speaks in this way necessarily has an idea of honesty; and if he pushes his obstinacy so far as to die for Atheism, he must have such a furious desire to be its Martyr, that he would be capable of exposing himself to the same torments, even if he were not an Atheist. (PD: OD 3, 118)[50]

What interests Bayle in this passage is twofold. Firstly, it is important for him to show that one can die for his faith but that one's can also die for one's non-faith: it is therefore possible to be a martyr of atheism. Hence, perseverance is not a criterion for the validation of a belief, just as moral virtue can be manifested in an individual independently of any religious conviction. Secondly, what this extract underlines is that Bayle was convinced that Islam had been the place of conservation of a form of philosophical free thought inherited from the Greek thinkers of Antiquity: according to Paul Rycaut, from whom the above extract is a mixture of quotations, Mahomet Effendi belonged to a sect (the "Muserins") to which Bayle sometimes refers following the English traveller and whose essential doctrine consisted of "absolutely denying divinity" (CPC: OD 3, 210). In Rycaut's book, the mention of Mahomet Efendi took place in a chapter devoted to the description of Muslim "sects", in the course of which he was careful to emphasise the existence of a growing number of Muslims who, even in the Sultan's entourage, believed in the divinity

---

[50] Some of the sentences in the quote are taken from Rycaut.

of Jesus Christ. Naturally, this is not what Bayle is interested in: what concerns him is the development, even in the imperial seraglio, of a sect preaching atheism and the identity of God and nature. In the same passage of the *Continuation des pensées diverses* aimed at demonstrating the inanity of the argument of the universal consensus in favour of the existence of God, Bayle follows up with another example: that of the existence, in Asia and China, of groups for which Spinoza's atheism is "the dogma". As we can see, while noting the existence of atheists in various cultural milieus, Bayle willingly associates these forms of atheism with Spinoza's thought. In short, he wanted to show that Spinoza's thought, associating God with nature, had always existed in some way, even if it was in the manner of a fire smouldering under the ashes. This is exactly what Bayle underlines in the very first remark of the article "Spinoza": "It has long been believed that the whole universe is but one substance and that God and the world are but one being" (DHC, Spinoza, A).

Bayle was convinced that a tradition of free thought had survived in Islam, leading to Spinoza and going back not only to Averroes but also to lesser-known figures in Arab-Muslim history. The article "Averroes" (which made its appearance in the 1702 edition of the *Dictionary*) would in fact deserve a study of its own. It should simply be noted that, as a sort of counterpart to the one devoted to Spinoza, it does not fail to make the link with the latter, in particular by proposing an argument against Averroes that also allows "one to invincibly refute Spinozism" and about which he refers in a note to his article devoted to the Jewish philosopher (DHC, Averroès, E).[51] In addition to Ibn Rushd, whom Bayle "read" only through the criticism offered by the *Commentaries* of the Jesuit College of Coimbra,[52] there are several figures whom Bayle identifies with a more or less coherent tradition of free thought. This is the case, for example, of the sect of "Ehl Eltahkik [ahl al-taḥqīq], or men of truth, people of certainty, who believe that there are for everything only the four elemens, which are God, which are man, which are all things",[53] or of the "Zindikites", heirs of the Sadducees, whose name they had taken over, and who maintained that "all that is seen [...], all that is in the world [...], all that has been created, is God" (DHC, Spinoza, A).[54]

This latter group, which we see Bayle link to a Jewish "sect" reputedly not believing in eternal life (DHC, Sadducéens), was in fact inspired by the Iranian revolt movement of the zanadiqa, "heretics" (zindīq-s being a generic term for "heretic" in Islam) which had arisen after the killing by the caliph al-Manṣūr of the Abbasid general Abū Muslim al-Khurāsānī (AD ca. 719–755). Of Iranian descent, Abū Muslim had contributed to the victory of the Abbasids over the Umayyads, before being put to death by the caliph for rather obscure reasons. If his death was

---

[51] Bayle explicitly refers to note N of the article "Spinoza".

[52] Carvalho 2013; Libera 2016, 119.

[53] These are probably the yāresān or Ahl-i ḥaqq in Persian (اهل حق), « people of the Truth », a Kurdish esoteric religion based on a syncretism between Shīʿīt mysticism, Zoroastrian doctrines and Manichean elements, close to the Bektashis, the Turkish Alevites, the Syrian Alawis or the Yezidis.

[54] The information was taken, once again, from Bespier 1677, 3, 548.

the signal for a Shīʿīste-inspired revolt in Iran, it is not certain that Abū Muslim intended to favour Shīʿīsm, and even less certain that he adhered to a particularly heterodox movement of thought. The fact remains that for Bayle, who here follows Erpenius's edition of al-Makin and Bespier's quotations referring to della Valle (DHC, Mahomet, A),[55] brings the monistic theory of the soul supposedly developed by this sect closer to the theses of Pythagoras, while showing the slight difference existing between their view of metempsychosis and that of the Ehl Eltahkik. The idea that the monistic theses of certain Greek philosophers condemned in the West had passed into Islam was easy to explain, according to Bayle:

> There are very wise people [like Paul Rycaut] who believe that the prodigious number of sects which are seen among the Turks, comes from the fact that there have been several people of different religions who have embraced Mahometism, either by interest or by force. The Greeks who did so, being from a country which was the school of the arts and sciences, mixed the ancient opinions of the philosophers with the reveries of the Alcoran, with which they were not too happy. (PD: OD 3, 59).

It was therefore the policy of conquest, but also the relative tolerance of the Muslims, which had been the cause of the survival of several Greek philosophical movements favouring theses considered heretical by the Christians who, for their part, had persecuted or even exterminated them. Bayle thus inaugurated a movement of rereading Islam within the free-thinking movement which was to flourish, a few years later, in the works of Mary Wortley-Montagu or John Toland (Léchot 2021, 293–296 and 393–410), and according to which, to quote Bayle, "the Arab philosophers followed Mahometism only in appearance and indeed mocked the Alcoran, because they found in it only those things which are contrary to reason" (DHC, Takkidin, A). The idea that Islam, behind the religious credulity of the masses, had always sheltered elite movements of thought with freer but hidden reflection, was thus to remain a constant among many thinkers of the Enlightenment whose interest in this form of "religio duplex" that Jan Assmann has highlighted is well known (Assmann 2014).[56]

3.6 Conclusion: Bayle, Lessing and the Islamic "Predilection" of Enlightened Philosophy

Far from the "objective" viewpoint that is sometimes attributed to him, Bayle's reading of Islam (at least as we have reconstructed it) actually highlights what prevented him from rising to the erudite level of Arabists such as d'Herbelot or Marracci, or what separated him from the orientalists who, like Adriaan Reland, were soon to advocate a more dispassionate and "objective" reading of the Islamic

---

[55] See Bespier 1677, 3, 392.

[56] One can find this distinction between the "credulity" of the masses and the clairvoyance of the philosophers in the worke of Vanini. See: Khayati 2009, 119.

religion.[57] His lack of knowledge of Arabic, his massive recourse to secondary literature (despite his reading of certain Arabic authors translated into Latin) as well as his insistence on the thesis of the Muhammadan imposture make Bayle an author who, all in all, is quite in line with the image of many other scholars who had then dealt with Islam without really being specialists in the subject.[58] Even when compared to the polemical and contextual approach of Prideaux, Bayle's view of Islam is much less well documented than that of the Anglican orthodox doctor.[59] Finally, Bayle's preoccupation with demonstrating several principles of his philosophical reflection through his developments devoted to the Islamic religion may leave the reader with the feeling of a kind of instrumentalization[60] or, in any case, of an interpretation oriented towards anthropological and philosophical concerns rather than purely scholarly ones. That said, even from a philosophical and Machiavellian perspective, Bayle's reading does not go as far as that of some freethinkers of the time who saw Muhammad as a false prophet but accorded him the status of a genius lawgiver—a tradition of reading that was to develop further during the Enlightenment and the Romantic period.[61]

However, and precisely because of this last point, the article "Mahomet" in the *Dictionnaire* and Bayle's other contributions about Islam reveal an important juncture in the history of the Western approach to Islam: that of the Islamic moment of early modern Western philosophy. With his article, Bayle offered a substantial and critical (i.e., historically and truth-conscious) summary of the many insights that had emerged in the century that was ending. As with Locke and Leibniz, Bayle's example underlines the extent to which Western philosophy could no longer ignore other monotheistic religions when it came to construct a system of religious philosophy. What I have described elsewhere as the "great decentring" (Léchot 2023b) began to bear its philosophical fruit under his pen: with the discovery of numerous Arab-Muslim sources and the increasingly sustained contacts between Europeans and Orientals, but also because of the development of critique as practised by Richard Simon, it became impossible to engage in philosophical reflection on the subject of religion without taking into account the "matters of fact" that concerned Islam. These factual elements, which Bayle identified above all with documents from the Arab-Muslim world, could no longer be avoided in his eyes—not least because of their "mass". Philosophy was therefore in a way summoned to take them into account in its conceptual developments and theoretical reformulations. As Georges Gusdorf has shown, the insistence on documents and facts signalled the emergence of a "new religious epistemology" (Gusdorf 1969, 2, 86–119). The study of Islam and Orientalist research thus contributed to the progressive triumph of this

---

[57] For a bibliographical synthesis on this question, see: Léchot 2023a.
[58] For example, some of Bayle's statements about Muhammad can be found in a sermon on Islam by his Genevan teacher Louis Tronchin. See Fatio 2015, 416–423.
[59] Mandelbrote 2022.
[60] Gobillot 2009, 71.
[61] Khayati 2009; Tolan 2019; Malcolm 2019.

"militant reason", which gave an account of reality to the detriment of metaphysical or polemical constructions increasingly associated by scholars with the coffin of Muhammad floating in the air (Gusdorf 1969, 3, 88). In this sense, Pierre Bayle constitutes a fundamental step in the intellectual and cultural evolution that saw Western thought move from the natural theology of the early seventeenth century to the natural history of religion such as Hume would outline in the following century (Lagrée 1991).

Bayle's influence on Enlightenment thought is there to confirm it. One of the best examples remains Lessing's. The German philosopher and playwright spent much of his early life reading Bayle and emphasising the importance of the most objective possible approach to Islam for philosophical reflection.[62] This "predilection" (Büttgen 2017, 22) for Islam was to culminate both in his writings on the occasion of the famous Quarrel of the Fragments and in the writing of the play that was to give him worldwide fame: *Nathan the Wise*. In all these works, we find the Baylean concern for a sincere and honest reading of Muslim sources as an inevitable prerequisite to any further reading of the religious phenomenon. Like Bayle, Lessing considered "honesty" (*Aufrichtigkeit*) towards historical and religious actors as an ethical sine qua non for any elaboration of a philosophy of religion.[63] Like Bayle, and with arguments partly reminiscent of his own, Lessing pointed to the inconsistency of Christians who never hesitate, in the name of their faith, to betray the precepts of their religion (the example of the Christian patriarch of Jerusalem in *Nathan* makes this particularly clear) while the Muslims, reputed to be lawless conquerors, know how to be tolerant and generous (as illustrated by the key figure of Saladin in Lessing's drama). And like Bayle, finally, what concerned Lessing was, in sum, how to escape the religious violence that the logic of confessional opposition inevitably engenders.

Naturally, Lessing develops his response within a philosophy of history that seems absent from Bayle's thinking, while his religious thought seems to seek to overcome the confessional confrontation in some way by the confrontation itself (but by changing its spirit), whereas Bayle seems to be content simply to denounce intolerance and religious coercion by means of his implacable logic and irony while calling for a universal tolerance that would encompass all religions and religious minorities but whose means of achievement seem hard to determine. For Lessing, in fact, society will be truly appeased on the religious level when the religions know how to peacefully confront each other in order to join, together, this natural religion of humanity that he calls for (Léchot 2021, 413–471). Yet, despite these differences, Lessing's thought, which draws on the roots of Islam to define its religious creed in the famous "parable of the rings" in *Nathan the Wise*, would not have been what it was if Bayle had not demonstrated that knowledge of Islam and meditation on its

---

[62] The German translation by Johann Christoph Gottsched and his wife Luise of the *Dictionnaire* was published from 1740 (see Espagne 2009). On Lessing and Bayle, see Nisbet 1978.

[63] In 1754, Lessing published his *Rettung des Hier. Cardanus*, which he conceived as a complement to Bayle's article (DHC, Cardan) and in which he already proposed an "honest" (*aufrichtig*) confrontation between the three great monotheistic religions (Léchot 2021, 424).

teachings in relation to those of Christianity was in some way the potential crucible of a profoundly renewed reflection in matters of philosophy of religion.

Bibliography

Anderson, Sonia. 1989. *An English consul in Turkey: Paul Rycaut at Smyrna, 1667–1678*. Oxford: Clarendon Press.
Assmann, Jan. 2014. *Religion Duplex. How the enlightenment reinvented Egyptian religion*. Oxford: Polity.
Atkinson, Geoffroy. 1924. *Les relations de voyage du XVIIe siècle et l'évolution des idées*. Paris: Champion.
Balagna Coustou, Josée. 1989. *Arabe et humanisme dans la France des dernier Valois*. Paris: Maisonneuve et Larose.
Bayle, Pierre. 1727–1731. *Oeuvres diverses de Mr Pierre Bayle, professeur en philosophie et en histoire à Rotterdam (OD)*, 4 vols. La Haye.
———. 1740. *Dictionnaire historique et critique* (DHC), 4 vols., 5th ed. Amsterdam/Leyde/La Haye/Utrecht: P. Brunnel et al.
———. *La Correspondance de Pierre Bayle*. http://bayle-correspondance.univ-st-etienne.fr/?lang=fr.
Bespier, Henri. 1677. Remarques curieuses sur la diversité des Dignitez & des Charges des Officiers du Grand-Seigneur, & sur les opinions differentes des Auteurs qui ont écrit de l'état de l'Empire Ottoman, le tout pour l'intelligence & l'éclaircissement de plusieurs choses. In *L'État present de l'empire ottoman ou sont compris les Mœurs, les Maximes, & la Politique des Turcs*, ed. Paul Rycaut, 3 vols. Rouen: J. Lucas.
Bevilacqua, Alexander. 2013. The Qur'an translations of Marracci and Sale. *Journal of the Warburg and Courtauld Institutes* 76: 93–130.
———. 2016. How to organise the orient: d'Herbelot and the 'Bibliothèque Orientale'. *Journal of the Warburg and Courtauld Institutes* 79: 213–261.
———. 2018. *The republic of Arabic letters. Islam and the European enlightenment*. Cambridge (MA) and. London: The Belknap Press of Harvard University Press.
Bibliander, Theodor. 1543. Machumetis Saracenorum principis ejusque successorum vitae, doctrina ac ipse Alcoran, quo velut authentico legum divinarum codice Agareni & Turcae, aliisque Christo adversantes populi reguntur, quae ante annos CCCC. In *Petrus abbas Cluniacensis per viros eruditos, ad fidei Christianae ac sanctae matris Ecclesiae propugnationem, ex Arabica lingua in Latinam transferri curavit*. Basel: J. Oporinus.
Bobzin, Hartmut. 2008. *Der Koran im Zeitalter der Reformation: Studien zur Frühgeschichte der Arabistik und Islamkunde in Europa*. Beyrouth and Würzburg: Ergon.
Booromeo, Elisabetta. 2007. *Voyageurs occidentaux dans l'Empire ottoman: inventaires des récits et études sur les itinéraires, les monuments remarqués et les populations rencontrées*. Paris: Maisonneuve et Larose.
Borgne, Émilie Le. 2022. Réflexions sur les langues dans les *Histoires orientales*. In Guillaume Postel (1510–1581), écrits et influence, eds. Paul-Victor Desarbres, Émilie Le Borgne, Frank Lestringant, Tristan Vigliano, 171–182. Paris: Sorbonne Université Presses.
Bost, Hubert. 1990. Histoire et critique de l'histoire chez Pierre Bayle. La *Critique générale de l'Histoire du Calvinisme de Mr. Maimbourg*, 1682–1683. *Revue d'histoire et de philosophie religieuses* 70: 69–108.
Burman, Thomas. 2007. *Reading the Qur'an in Latin Christendom, 1140-1560*, Philadelphia: University of Pennsylvania Press.
Busbecq, Ogier Ghislain de. 2010. *Les Lettres turques*, trad. Dominique Arrighi. Paris: Champion.

Büttgen, Philippe. 2017. Confession et migration: l'islam des Lumières. In *Gotthold Ephraim Lessing. Adam Neuser (1774)*, trad. Philippe Büttgen, 5–78. Paris: Demopolis.
Carvalho, Mário Santiago de. 2013. Pierre Bayle et la critique d'Averroès à Coimbra. Deux épisodes de l'histoire de la réception d'Averroès. *Revista Filosófica de Coimbra* 44: 417–432.
Champion, Justin. 2010. 'I remember a Mahometan story of Ahmed Ben Edris': Freethinking uses of Islam from Stubbe to Toland. *Al-Quantara* 31: 443–480.
Chardin, Jean. 1686. *Journal du voyage du Chevalier Chardin en Perse & aux Indes orientales, par la Mer Noire & par la Colchide. Qui contient le Voyage de Paris à Isfahan*. Amsterdam: A. Wolfgangh.
Charnley, Joy. 1998. *Pierre Bayle reader of travel literature*. Berne: P. Lang.
D'Herbelot de Molainville, Barthélemy. 1697. *Bibliothèque orientale, ou Dictionnaire universel contenant generalement Tout ce qui regarde la connoissance des Peuples de l'Orient*. Paris: Compagnie des Libraires.
Daireaux, Luc. 2010. *Réduire les huguenots: protestants et pouvoirs en Normandie au XVIIe siècle*. Paris: Champion.
Dandini, Jérôme. 1675. *Voyage du Mont Liban*, traduit de l'Italien, trad. Richard Simon. Paris: L. Billaine.
Daniel, Norman. 1960. *Islam ans the West. The making of an image*. Edinburgh: The University Press.
Dew, Nicholas. 2004. The order of oriental knowledge: The making of d'Herbelot's *Bibliothèque Orientale*. In *Debating world literature*, ed. Christopher Prendergast, 233–252. London/New York: Verso.
———. 2009. *Orientalism in Louis XIV's France*. Oxford: Oxford U.P.
Dyâb, Hanna. 2015. *D'Alep à Paris. Les pérégrinations d'un jeune Syrien au temps de Louis XIV*, eds and trad. Paul Fahmé-Thiéry, Bernard Heyberger and Jérôme Lentin. Paris: Acte Sud.
Erpenius, Thomas. 1617. *Sūrat Yūsuf wa-tahajjī al-'Arabi. Historia Josephi Patriarchae, ex Alcorano Arabice*. Leiden: Ex Typographia Erpeniana.
———. 1625. *Tārīḫ al-muslimīn min ṣāḥib sharī'at al-islām Abī al-Qāsim Muḥammad ilá al-dawla al-atābakīya id est, Historia saracenica, qua res gestae muslimorum inde a Muhammede primo imperij et religionis muslimicae auctore, usque ad initium imperij Atabacaei per XLIX imperatorum successionem fidelissime explicantur, insertis etiam passim christianorum rebus in Orientis potissimum ecclesijs eodem tempore gestis, arabice olim exarata a Georgio Elmacino*. Leiden: Ex Typographia Erpeniana linguarum orientalium.
Espagne, Michel. 2009. Lessing et les hérétiques. *Revue Germanique Internationale* 9: 133–145.
Fatio, Olivier. 2015. *Louis Tronchin. Une transition calvinienne*. Paris: Garnier.
Gallien, Claire. 2015. Edward Pococke et l'orientalisme anglais du XVIIe siècle: passeurs, transferts et transitions. *Dix-septième siècle* 268: 443–458.
Girard, Aurélien. 2011. *Le christianisme oriental (XVIIe-XVIIIe siècles): essor de l'orientalisme catholique en Europe et construction des identités confessionnelles au Proche-Orient*. Paris: École pratique des Hautes Études (PhD thesis).
Glei, Reinhold F., and Roberto Tottoli. 2009. *Ludovico Marracci at work. The evolution of his Latin translation of the Qur'an in the light of his newly discovered manuscripts*. Wiesbaden: Harrassowitz.
Gobillot, Geneviève. 2009. Les approches de l'islam au XVIIe siècle à travers la science et la philosophie. In *L'Islam visto da Occidente. Cultura e religione del Seicento europeo di fronte all'Islam*, ed. Bernard Heyberger, Mercedes Garcia-Arenal, Emanuele Colombo, and Paola Vismara, 39–74. Milan: Marietti 1820.
Gusdorf, Georges. 1969. *La révolution galiléenne*, 2 vols. Paris: Payot.
Hamilton, Alastair, and Francis Richard. 2004. *André Du Ryer and oriental studies in Sevennheenth-Centyr France*. London: The Arcadian Library.
Hanne, Olivier. 2019. *L'Alcoran. Comment l'Europe à découvert le Coran*. Paris: Belin.
Heyberger, Bernard. 1994. *Les chrétiens du Proche-Orient au temps de la Réforme catholique (Syrie, Liban, Palestine, XVIIe-XVIIIe siècle)*. Rome: École française de Rome.

———. 2015. L'Orient et l'islam dans l'érudition européenne du XVIIe siècle. *Dix-septième siècle* 268: 495–508.
Jaquin, Frédéric. 2010. *Le voyage en Perse au XVIIe siècle*. Paris: Belin.
Jurieu, Pierre. 1687. *Des Droits des deux souverains en matière de religion, la conscience et le Prince. Pour détruire le dogme de l'indifference des Religions & la tolerance Universelle: contre un Livre intitulé Commentaire philosophique*. Rotterdam: H. de Graef.
Khayati, Loubna. 2009. Le statut de l'islam dans la pensée libertine du premier XVIIe siècle. In *L'Islam visto da Occidente. Cultura e religione del Seicento europeo di fronte all'Islam*, ed. Bernard Heyberger, Mercedes Garcia-Arenal, Emanuele Colombo, and Paola Vismara, 109–133. Milan: Marietti 1820.
Koningsveld, Pieter Sjoerd Van. 1996. The Islamic image of Paul and the origin of the gospel of Barnabas. *Jerusalem Studies in Arabic and Islam* 20: 200–228.
Labrousse, Élisabeth. 1957. La méthode critique chez Pierre Bayle et l'Histoire. *Revue Internationale de Philosophie* 11 (42): 450–466.
———. 1996. *Pierre Bayle: hétérodoxie et rigorisme*. Paris: Albin Michel.
Lagrée, Jacqueline. 1991. *La religion naturelle*. Paris: Presses Universitaires de France.
Larzul, Sylvette. 2009. Les premières traductions françaises du Coran (XVIIe-XIXe siècle). *Archives de sciences sociales des religions* 147: 3–42.
Laurens, Henry. 1978. *Aux sources de l'orientalisme. La Bibliothèque orientale de Barthélemi d'Herbelot*. Paris: Maisonneuve et Larose.
Léchot, Pierre-Olivier. 2021. *Luther et Mahomet. Le protestantisme d'Europe occidentale devant l'Islam (XVIe-XVIIe siècle)*. Paris: Éditions du Cerf.
———. 2022. 'Apostat, transgresseur de la Loi et intrus dans le vrai christianisme'. Racines confessionnelles et transculturelles de l'antipaulinisme des Lumières. *Études théologiques et religieuses* 97: 129–157.
———. 2023a. Quand un calviniste explorait le jihād. Adriaan Reland (1676-1718) et les motivations de l'érudition orientaliste à l'aube des Lumières. In *Chanter l'histoire en réveillant les sources. Mélanges d'histoire(s) offerts au professeur Jean-Daniel Morerod*, ed. Lionel Bartolini, Grégoire Oguey, and Isaline Deléderray-Oguey, 323–362. Neuchâtel: Alphil-Presses Universitaires de Suisse.
———. 2023b. Déjudaïser la Bible? Les hésitations de l'érudition protestante de l'Âge classique devant la littérature rabbinique. *Études théologiques et religieuses* 98: 43–70.
Lestringant, Frank. 1985. Guillaume Postel et l''obsession' turque. In *Guillaume Postel (1581-1981), Actes du Colloque International d'Avranches, 5-9 septembre 1981*, ed. Guy Tredanel, 265–298. Paris: Éditions de la Maisnie.
Libera, Alain de. 2016. Averroès, le philosophe à barbe. *Association Française des Acteurs de l'Éducation* 151: 115–122.
Loop, Jan. 2013. *Johann Heinrich Hottinger. Arabic and Islamic studies in the seventeenth century*. Oxford: Oxford U.P.
Malcolm, Noel. 2019. *Useful enemies. Islam and the ottoman empire in Western political thought, 1450-1750*. Oxford: Oxford U.P.
Mandelbrote, Scott. 2022. The significance of historical Judaism and the career of Humphrey Prideaux. In *The Mishnaic moment. Jewish law among Jews and Christians in early modern Europe*, ed. Piet van Boxel, Kirsten MacFarlane, and Joanna Weinberg, 255–277. Oxford: Oxford UP.
Marracci, Ludovico. 1691a. *Prodromus ad refutationem Alcorani. In quo per quatuor praecipuas verae Religionis notas Mahumetanae Sectae falsitas ostenditur: Christianae Religionis veritas comprobatur*. Rome: Typis Sacrae Congregationis de Propaganda Fide.
———. 1691b. *Prodromus ad refutationem Alcorani. In quo per quatuor praecipuas verae Religionis notas Mahumetanae Sectae falsitas ostenditur: Christianae Religionis veritas comprobatur*. Rome: Typis Sacrae Congregationis de Propaganda Fide.
———. 1698. *Refutatio Alcorani, in qua ad Mahumetanicae superstitionis radicem securis apponitur; & Mahumetus ipse gladio suo jugulatur*. Padua: Ex Typographia Seminarii.

Matar, Nabil. 1991. John Locke and the Turbanned nations. *Journal of Islamic Studies* 2: 67–77.

———. 2015. England and religious plurality: Henry Stubbe, John Locke and Islam. *Studies in Church History* 51: 181–203.

Mills, Simon. 2020. *A commerce of knowledge. Trade, religion, and scholarship between England and the ottoman empire, c. 1600–1760*. Oxford: Oxford U.P.

Morrow, John Andrew. 2015. *Six covenants of the prophet Muhammad with the Christians of his time. The Primary Documents*. New York: Covenant Press.

———. 2016. Johann Georg Nissel. In *Christian-Muslim relations. A bibliographical history*. Vol. 8. *Northern and Eastern Europe (1600–1700)*, ed. David Thomas and John Chesworth, 608–617. Leiden: Brill.

Nisbet, Hugh B. 1978. *Lessing* and Pierre *Bayle*. In *Tradition and creation: Essays in honour of Elizabeth Mary Wilkinson*, ed. C.P. Magill, B.A. Rowley, and C.J. Smith, 13–29. Leeds: Maney.

Ouardi, Hela. 2017. *Les derniers jours de Muhammad*. Paris: Albin Michel.

Postel, Guillaume. 1560. *De la Republique des Turcz*. Poitiers: E. de Marnef.

———. 1575. *Des histoires orientales et principalement des Turkes ou Turchikes et Szchitiques ou Tartaresques et aultres qui en sont descendues, Œuvre pour la tierce fois augmentée*. Paris: H. de Marnef and G. Cavellat.

Prideaux, Humphrey. 1698. *La vie de Mahomet, où l'on découvre amplement la Vérité de l'Imposture, trad. Daniel de Larroque*. Amsterdam: G. Gallet.

Richard, Francis. 1995. *Raphaël du Mans, missionnaire en Perse au XVIIe siècle*. Paris: L'Harmattan.

Richard, Robert, and Denis Vatinel. 1981. Le Consistoire de l'Eglise réformée du Havre au XVIIe siècle: les pasteurs (étude sociale). *Bulletin de la Société de l'Histoire du Protestantisme français* 127: 1–77.

Rizzi, Massimo. 2007. *Le prime traduzioni del Corano in Italia: contesto storico e attitudine dei traduttori. Ludovico Marracci (1612–1700) e la lettura critica del commentario coranico di al-Zama'šarī (1075–1144)*. Turin: L'Harmattan Italia.

Rycaut, Paul. 1668. *The present state of the Ottoman Empire. Containing the maxims of the Turkish Politie, the most material points of the Mahometan religion, their sects and heresies, their convenents and religious votaries, their military discipline with an exact computation of their forces both by land and sea*. London: J. Starkey and H. Brome.

———. 1670. *Histoire de l'État présent de l'Empire ottoman: contenant les maximes politiques des Turcs; les principaux Points de la Religion Mahometane, ses sectes, ses Herésies, & ses diverses sortes de Religieux; leur Discipline Militaire, avec une supputation exacte de leurs Forces par mer & par terre, & du revenu de l'Etat, trad. Pierre Briot*. Paris: S. Mabre-Cramoisy.

———. 1677. *L'État present de l'Empire ottoman ou Sont compris les Mœurs, les Maximes, & la Politique des Turcs; leurs manieres de gouverner; leur discipline Militaire; leur Religion; leurs Mariages; leurs forces par Mer & par Terre; & comment le Grand-Seigneur se maintient dans l'éclat & la gloire, & se fait craindre, trad. Henri Bespier*. Rouen: Jacques Lucas.

Ryer, André du. 1647. *L'Alcoran de Mahomet translaté d'arabe en françois*. Paris: A. de Sommaville.

Said, Edward W. 2003. *L'orientalisme. L'Orient créé par l'Occident, trad. Catherine Malamoud*. Paris: Éditions du Seuil.

Schino, Anna Lisa. 2020. *Batailles libertines. La vie et l'œuvre de Gabriel Naudé*. Paris: Champion.

Simon, Richard. 1684. *Histoire critique de la Creance et des Coutumes des nations du Levant, publiée par le Sr. De Moni*. Francfort [=Rotterdam]: F. Arnaud.

———. 1983. In *Additions aux Recherches curieuses sur la diversité des langues et religions d'Edward Brerewood*, ed. Jacques Le Brun and John D. Woodbridge. Paris: Pressues universitaires de France.

Sionita, Gabriel. 1630. *Testamentum et pactiones initiae inter Mohamedem et Christianae fidei cultores*. Paris: Vitré.

Stroumsa, Guy. 2010. *A new science. The discovery of religion in the age of reason*. Cambridge, MA/London: Harvard University Press.

Tavernier, Jean-Baptiste. 1676. *Six voyages de Jean-Baptiste Tavernier qu'il a faits en Turquie, en Perse et en Indes*, 2 vols. Paris: G. Clouziers and C. Babin.

Thomson, Ann. 2017. Henri Bespier. In *Christian-Muslim relations. A bibliographical history. Vol. 9. Western and Southern Europe (1600–1700)*, ed. David Thomas and John Chesworth, 561–564. Leiden: Brill.

Tinguely, Frédéric. 2000. *L'écriture du Levant à la Renaissance. Enquête sur les voyageurs français dans l'empire de Soliman le Magnifique*. Genève: Droz.

———. 2022. Le temps du jugement dans *La République des Turcs*. In *Guillaume Postel (1510–1581), écrits et influence*, ed. Paul-Victor Desarbres, Émilie Le Borgne, Frank Lestringant, and Tristan Vigliano, 171–182. Paris: Sorbonne Université Presses.

Tolan, John. 2002. *Saracens, Islam in the medieval European imagination*. New York: Columbia U.P.

———. 2019. *Faces of Muḥammad: Western perceptions of the prophet of Islam from the middle ages to today*. Princeton/Oxford: Princeton U.P.

Toomer, Gerald J. 1996. *Eastern Wisedome and learning: The study of Arabic in seventeenth-century England*. Oxford: Clarendon Press.

Torabi, Dominique. 1992. La Perse de Barthélemy d'Herbelot. *Luqmân* 8: 43–58.

van Boxel, Piet, Kirsten MacFarlane, and Joanna Weinberg, eds. 2022. *The Mishnaic moment. Jewish law among Jews and Christians in early modern Europe*. Oxford: Oxford UP.

Van der Cruysse, Dirk. 1998. *Chardin le Persan*. Paris: Fayard.

———. 2002. *Le noble désir de courir le Monde. Voyages en Asie au XVIIe siècle*. Paris: Fayard.

Van der Lugt, Mara. 2016. The body of Mahomet: Pierre Bayle on war, sex, and Islam. *Journal of the History of Ideas* 78: 27–50.

Varani, Giovanna. 2008. Leibniz und der Islam: Die Betrachtung des Korans als erster Ansatz zu einer Kulturbegegnung im 17. Und 18. Jahrhundert. *Studia Leibnitiana* 40: 48–71.

Vigliano, Tristan, and Mouhamadoul Khaly Wele. 2021. Le droit de traduire le Coran: réflexions sur la version française d'André du Ryer. In *Discours et stratégies d'altérité. Regards et analyses croisés*, ed. Ali Mostfa, 115–126. Paris: L'Harmattan.

Vrolijk, Arnoud. 2009. The prince of Arabists and his many errors: Thomas Erpenius's image of Joseph Scaliger and the edition of the *Proverbia Arabica* (1614). *Journal of the Warburg and Courtauld Institutes* 72: 143–168.

Vrolijk, Arnoud, and Richard van Leeuwen. 2014. *Arabic studies in The Netherlands. A short history in portraits, 1580–1950*. Leiden: Brill.

Vrolijk, Arnoud, and Joanna Weinberg. 2020. Thomas Erpenius, oriental scholarship and the art of persuasion. In *Scholarship between Europe and the Levant. Essays in honour of Alastair Hamilton*, ed. Jan Loop and Jill Kraye, 34–59. Leiden: Brill.

Whelan, Ruth. 1989. *The anatomy of superstition: A study of the historical theory and practice of Pierre Bayle*. Oxford: Voltaire Foundation.

Williams, Benjamin. 2022. Bringing Maimonides to Oxford: Edward Pococke, the Mishnah, and the *Porta Mosis*. In *The Mishnaic moment. Jewish law among Jews and Christians in early modern Europe*, ed. Piet van Boxel, Kirsten MacFarlane, and Joanna Weinberg, 156–176. Oxford: Oxford UP.

Chapter 4
Bayle: Confucianism and China

Marta García-Alonso

Abstract China plays an essential role in Bayle's philosophical work. He uses China to demonstrate that there is no necessary connection between religion and a successful society. Politics and morality do not require a religious root. Besides, the atheism of the Chinese is not merely a negative absence of belief, like that of the Americans or Africans. China is not simply a backdrop for the creation of the philosophical myth of atheism but a model for political tolerance. This is supported by the equation between Confucianism and civic religion, an idea that may have come from the works of the Jesuits. China embodies political tolerance by not imposing any obligations to adhere to a particular doctrine or participate in a specific religious ceremony. This type of toleration is a religious-political model where beliefs and rituals, the two fundamental elements of any religion, are not mandatory for the entire population.

Keywords China · Confucianism · Bayle · Political tolerance · Atheism

---

This paper was made possible by the research project *Contra la ignorancia y la superstición: las propuestas ilustradas de Bayle y Feijoo* (PID2019-104254GB-100) financed by the Spanish Ministry of Education and Science. I am very grateful to J. C. Laursen for a careful reading of the manuscript and helpful suggestions and comments.

M. García-Alonso (✉)
Departamento de Filosofía Moral y Política, UNED, Madrid, Spain
e-mail: mgalonso@fsof.uned.es

© The Author(s), under exclusive license to Springer Nature Switzerland AG 2024
M. García-Alonso, J. C. Laursen (eds.), *The Importance of Non-Christian Religions in the Philosophy of Pierre Bayle*, International Archives of the History of Ideas Archives internationales d'histoire des idées 251, https://doi.org/10.1007/978-3-031-64865-6_4

4.1 Introduction: The Jesuit Rites Controversy

The context in which a philosophical discussion takes place is essential for its comprehension. In the case of Bayle's theses on China, the debate surrounding the Jesuit missions in the East (Gernet 1982; Laven 2011; Mungello 2009; Pinot 1932; Zoli 1989) enables our understanding. The Rites Controversy encompassed two interconnected theological questions. First, it addressed the feasibility of translating Christian concepts into Chinese and equating the Chinese sky god with the Christian god. Second, it debated the extent to which newly baptized converts can continue to participate in Confucian funeral rites, a matter of great significance as many converts were members of the ruling elite who had adopted Confucianism as their political ideology since 136 CE.

When missionaries journey to unfamiliar lands, they often face the challenge of translating key Christian beliefs into the local language. They must also decide how to interpret the ceremonies they encounter, either to make them compatible with Christianity or to reject them as superstitions. Determining which aspects of Chinese culture can be accepted by the Christian West and which must be rejected was a delicate and complex issue, as the conversion of a people to Christianity hung in the balance. One controversial thesis held by Jesuit missionaries was that Confucius was a sage rather than a god or saint. In China, Confucius was revered as a master, but prayers were not addressed to him. The ceremonies surrounding him were simply meant to show respect for the dead. Therefore, the wooden tablets used in these ceremonies did not represent the souls of the deceased. Confucianism did not view the dead as mediators between God and man or as sources of favours, as was the case with Christian saints. On the other hand, the sacrifices that Emperors offered to heaven were made to the Lord and creator of Heaven and Earth and all that exists (*Shangdi*), not to the physical heaven. Therefore, missionaries argued that the terms Lord of Heaven (*Tianzhu*), Heaven (*Tian*), and Supreme Emperor or Lord of Heaven (*Shangdi*) were interchangeable. However, Charles Maigrot (1652–1730), the papal vicar in China, and other critics of the Jesuits argued that this translation confused and undermined Christian principles. They also argued that the ceremonies honouring ancestors and the rites associated with Confucius were pure paganism.

As expected, the disagreement escalated, and Pope Clement XI was forced to take a stance in 1704 by prohibiting the use of the terms *Tian* and *Shangdi* to refer to the Christian God. However, this papal directive had little practical impact on the eastern missions, where the Jesuits continued to hold their own beliefs on the matter. The pope had no choice but to reaffirm his position with the Bull *Ex illa die* (1715), in which he declared that the West only refers to the creator of the universe as Deus and therefore the terms *Tian* or *Shangdi* should not be used to refer to him. Similarly, the bull stated that ceremonies honouring Confucius and ancestors are pagan cults that are forbidden to converts, whether they perform or attend as spectators. According to this prohibition, Chinese government officials who have converted to Catholicism are not allowed to participate in Confucian ceremonies. No Catholic is permitted to venerate their ancestors either in private (in family temples) or in public (in cemeteries or at funerals). This prohibition created a significant obstacle for the Jesuits' mission in China and their goal of influencing government officials. This obstacle only grew when the Pope reminded the Chinese Emperor that he should not interfere with a ban on converts, even if they were his subjects.

But the discussion on how to interpret or accept Chinese ceremonies was far from over, and it would engage the minds of European intellectuals for over a century and a half. Seven popes, two Chinese emperors, the kings of Portugal, Spain, and France, Louis XIV's confessor François d'Aix de la Chaise, the Sorbonne University, Jansenists such as Blaise Pascal and Antoine Arnauld, Calvinist theologians like Pierre Jurieu, Catholics like François Fénelon and Jacques-Bénigne Bossuet, and philosophers like Bayle and Leibniz were all involved in this dispute. As we can see, both the opponents and defenders of the Jesuit thesis were numerous and influential and came from various religious backgrounds. The climax of the controversy came with the denunciation by the Faculty of Theology at the Sorbonne of the theses of Louis le Comte (1655–1728) in his works *Nouveaux Mémoires sur l'État présent de la Chine* (1696) and *Lettre à Monseigneur le Duc du Mayte sur les Cérémonies de la Chine* (1700), as well as the assertions of Charles Le Gobien (1653–1708) in his *Histoire de l'Édit de l'Empereur de la Chine en faveur de la religion chrétienne* (1698), both of whom Bayle quotes extensively in his own works. The debate finally came to an end in 1742 when Pope Benedict XIV issued the bull *Ex quo singulari*, condemning Confucianism and forbidding Christians from participating in its rituals.

In this context, Bayle writes on the nature of paganism, engages in debates on the concept of a universal natural religion, and, most significantly, suggests the possibility of a society governed by atheistic philosophers.

4.2 Bayle's Sources on China

In 1683, when Bayle wrote the *Pensées diverses*, he was not yet aware of or interested in the case of China. At that time, he himself acknowledged that there was no evidence of any atheist nation and therefore no proof that atheists could exhibit any kind of moral behaviour (PD: OD III, 81; 93). However, this changed after he became familiar with travel accounts between 1684 and 1687 while writing the *Nouvelles de la république des lettres*, as these books provided Bayle with a wealth of information on Chinese customs and religious beliefs (Charnley 1990; Bayerl 2012). From 1685 onwards, philosophers used the knowledge gained from Asia to discuss the religious, moral, and political ideas of their contemporary context. From 1760 onwards, the focus shifted from criticism to the construction of the new social science of the physiocrats, with China at the centre of both movements (Pinot 1932, 10 ff). In both cases, the material provided by Christian missionaries was essential. They were responsible for identifying and selecting in Eastern culture those elements that legitimated the socio-political order, separating behaviours that could be defined as superstitious or idolatrous and therefore subject to refutation and elimination. As idolatry and superstition could be disguised in any type of practice, the most detailed analysis possible of the customs of these places and their inhabitants was necessary to combat it. For this reason, the missionaries' interest extended to all areas, including religious ceremonies, the list and function of their gods, astrology, morals, philosophy, ethics, botany, medicine, types of food and drink, music and theatre, laws, gardens, burial rites, and so on. In this task of compilation, the assistance of new converts was no less important.

One of the problems that missionaries encounter is how to interpret Confucian rites that involve the faithful's relationship with the spiritual tablets of their ancestors –these ceremonies involve the reverential treatment of the tablets, the burning of incense and money in their honour, the offering of food and drink, and even the sacrifice of animals when honouring Confucius (Minamiki 1985, 3–11). The question arises as to whether these ceremonies constitute a natural religion, a civic cult, or simple idolatry. Additionally, it is unclear if the beliefs of the elite can be distinguished from those of the masses in the performance of these rituals, and what significance can be attributed to any potential difference. Additionally, the role of political representatives in such worship and how their participation may affect their constituents is a matter of concern. This discussion was mentioned explicitly in 1704 when Bayle wrote his *Continuation des pensées diverses à l'occasion de la Comète*.[1] As we will see, these are the same concerns of Bayle, who was familiar with the controversy over rituals as he acknowledges in a letter to Minutoli dated 1692.

It has been a lengthy period since we last encountered the works of Mr. Arnaud, as you mentioned to me. It is in relation to this extended duration that I expressed my surprise to you. I have just learned that he has not been indolent, but rather has been occupied with the production of the sixth volume of the *Morale Pratique*, which has recently been published. This volume is even more intriguing than the two preceding ones, as it addresses the worship of Confucius in China and the dispute between the Dominicans and Jesuits over whether this worship constitutes a religion or a civil practice. The Jesuits have argued through their shrewdness that it is the latter and have explored how one can participate in it without committing idolatry.[2]

---

[1] Bayle's works are cited using the initials of the work, followed by the volume in which it is found in the *Oeuvres diverses* (OD) and the page. PD: *Pensées diverses écrites à un Docteur de Sorbonne, à l'occasion de la comète* (1683); NRL: *Nouvelles de la république des lettres* (1684–1687); FC: France toute catholique; CP: *Commentaire philosophique sur ces paroles de Jésus-Christ: contrain-les d'entrer* (1686); Supplément: *Supplément au Commentaire Philosophique* (1688); APD: *Adition aux Pensées diverses sur les Cometes* (1694); RQP: *Réponse aux question d'un Provincial* (1703–1707). We cite the *Dictionnaire*: DHC, title of article, remark number. Also, I have use the *Correspondance de Pierre Bayle* available online: http://bayle-correspondance.univ-st-etienne.fr/?lang=fr (access January 2023). When quoting in the body of the text from an English edition of Bayle's works, we will indicate the location of the paragraph in both editions.

[2] Lettre 895: Pierre Bayle à Vincent Minutoli, 11 noviembre 1692:
http://bayle-correspondance.univ-st-etienne.fr/?Lettre-895-Pierre-Bayle-a-Vincent&lang=fr

Although the Jesuits are not the only missionaries present in Asia—the Augustinians also enjoyed high regard among dignitaries in Persia, India, China, and Japan—they are the ones who feature prominently in Bayle's work. He recognizes in the Jesuits a scientific and cultural education that goes beyond the theological knowledge or catechism that they go to preach in those lands (NRL: OD I, 650). Their biographies, texts, and ideas are frequently mentioned in his periodical, such as the *Voyage de Siam des Pères Jesuites* (1686) by Guy Tachard, whom he also quotes in his *Dictionnaire* (DHC, Brachmanes; Sommona-Codom). The discussion of Le Comte's theses can be found in numerous pages of Bayle's work, but particularly in the *Continuation des Pensées diverses* (CPD: OD III, §XXVII-XIX). Similarly, in *Nouvelles de la république des lettres*, Ferdinand Verbiest, Phillippe Couplet, Martino Martini, Giovanni Battista Riccioli, Boymus, Polonois, and Nicolas Trigault are mentioned. Of course, he also quotes Cosimo Ruggeri (RQP: OD III, 13). The *Dictionnaire* includes several references to the humanist and orientalist Guillaume Postel –who was closely associated with the Society of Jesus– and his work *Des Histoires Orientales* (1575) (DHC, Albufeda B; Amphiaraus, n. 89; Chederles B, n.3). Bayle also cites Isaac Vossius's *Variorum Observationum Liber* (1685), in which Vossius discusses the peaceful nature of the Chinese and their inclination towards the study of philosophy, to the point that their political advisors are philosophers (NRL: OD I, 214). Bayle also references Adam Schall von Bell, who wrote 52 volumes on astronomy and was granted various honors and prebends by the Chinese Emperor, and Matteo Ricci, who was a pensioner of the emperor in Beijing (NRL: OD I, 663). Bayle also mentions the Sicilian Jesuit Nicolò Longobardi, who wrote *De Confucius ejusque doctrina tractatus* (1622–1625) promoting the theory that the Chinese had no concept of God, angels, or the soul, in contrast to Ricci. The discussions on the theses of Le Comte and Longobardi regarding China can be found in *Réponse aux questions d'un provincial* (RQP: OD III, 926–27). Longobardi's work, despite being prohibited, was preserved by Domingo Navarrete in a Spanish version in his *Tratados históricos, políticos, ethicos de la monarchia de China* (Busquets 2015). From these sources, Bayle concludes that Ricci is one of the originators of pantheistic atheism due to his equating of the material heaven with God (DHC, Maldonat, L).

Bayle's assumption of the equation between Confucianism and civic religion may have come from the works of Ricci and his followers. Matteo Ricci is the author of *De Christiana expediciones apud Sinas* (1615) which was translated and published by Belgian Jesuit Nicolas Trigault. This text provides a comprehensive overview of various aspects of life in China, including geography, demography, agriculture, diet, arts, technology, sciences, rituals, and religions. The text was highly successful, with several Latin (1616, 1617, 1623, 1684) and three French reprints at Lyon (1616, 1617, 1618), as well as translations into German, Italian, English, and Spanish. It is considered to be the most influential book on China of its time (Mungello 1985, 48). Ricci believed that there was no conflict between Christian theology and Chinese political thought, as Confucianism promotes harmony and hierarchy and emphasizes moral self-cultivation among the elites destined for government. However, Ricci's text advances the notion that Confucianism should be equated with secular and civic philosophy, which presupposes a division between the political and the religious, and the rational and the emotional, a concept that is characteristic of Western thought, but alien to the Chinese tradition. According to Gernet (1982, 63), every doctrine in Chinese culture has political implications, and the idea of a true and unique religion is meaningless.

In his work *Chinese Dialogue Tianzhu shiyi or True Meaning of the Lord of Heaven*, published in 1603, Ricci argues vehemently that the introduction of Buddhism and Taoism into China transformed Confucianism into a superstitious practice. He argued that only by forming an alliance between Christianity and the true Chinese culture can a new hybrid religion emerge, one that will eradicate Buddhist idolatry and revitalize Confucianism.[3] Thus, Ricci interpreted the Confucian tradition not in isolation, but always in the context of its potential adaptation and integration with Christian doctrine. It was usual in China to accept religions based on their adherence to tradition, which gave them legitimacy through their contribution to public morality, social order, and civil peace.

What Ricci was particularly interested in was strengthening the notion that the classical Chinese texts embody the concept of the true God. For this reason, he highlighted the tragedy of China's deviation from the worship of the Lord of Heaven in favour of Buddhist and Taoist idols. Hence, Ricci's work not only aimed at promoting a reconciliation between Christianity and Confucianism, but also at criticizing what he perceived as the paganism and atheism inherent in Buddhism, as noted by Gernet (1982, 37) and Mungello (1985, 64). According to Hsia, Ricci's work could even be considered an anti-Buddhist machine of war (2010). It was imperative for Jesuits to challenge the spiritual and religious beliefs of Buddhism and Taoism in order to establish Christianity as the only true religion in the Confucian worldview. By convincing the Chinese that Confucianism had been corrupted by the incorporation of Buddhist ideas, Jesuits could fulfil their objective and impose monotheism. And if that required manipulating the interpretation of the Chinese classics, so be it. In a letter written to the General of the Jesuits in 1604, Ricci writes:

> Given that the scholars who hold power in China take great offense if we criticize this principle (of *Taiji*), we have been more focused on challenging their explanation of the principle rather than the principle itself. If, in the end, they come to understand that *Taiji* is the first substantial, intelligent, and infinite principle, then we would agree that it is indeed God and nothing else.[4]

Therefore, to draw the Chinese towards Christian theology, any method was considered acceptable, including presenting Christianity as a rational philosophy compatible with the sciences. As a result, the Jesuits who travelled to China presented themselves as mathematicians, astronomers, musicians, or doctors, and were often perceived as scientists. This manoeuvring was replicated for the French Jesuits, as Bayle wrote in 1686 in *La France toute catholique*:

---

[3] In this volume, Fernando Bahr details how the feudal lords of Japan provided support to the Jesuits with the aim of curtailing the growth of Buddhism within their territories.

[4] Ricci in Gernet 1982, 41–42. Translation mine.

And there is no doubt that the same laws of humanity oblige an honorable man to inform the emperor of China what has just happened in France, so that he can take his measures to receive suitably the missionaries whom the king has just sent into that country on the footsteps of some great mathematicians. One is conscience-bound to warn that emperor that those people, who begin by asking merely to be tolerated, have as their real goal to become the masters and then compel everybody with a knife to their throats to be baptized without heed of any oath, edict, or treaty made for the safekeeping of the old religion. Now suppose that these missionaries persuade some Chinese to become Christians and that with these converts they undertake to force the rest, believing they are obliged by the parable 'compel them to come in'; suppose also that the emperor of China strongly opposes them and it comes to a declaration of war; he cannot feel sure about any agreement reached with his Christian subjects since, as soon as the occasion presents itself, the missionaries will say to the Chinese Christians that they have only promised the emperor to remain peaceful without forcing other Chinese to convert for a limited time, with conditions and subject to the law of the Church and the Gospel. For all intents and purposes that would be no worse than the revocation of the Edict of Nantes (FC: OD II, 350–1/ 2014, 27).

The intellectual converts, such as Yang Tingyun (1557–1627), Li Zhizao (1565–1630), and Xu Guangqi (1562–1633)—the highest-ranking official in the Ming dynasty–, collectively known as the *Three Pillars* of the Christian Church, were drawn to Ricci specifically due to his scientific expertise. These individuals were responsible for translating significant works of European science and philosophy into Chinese. Xu was the first to be baptised, followed by Li Zhizao and Yang Tingyun. Of the three, Xu Guangqi was particularly influential in shaping Ricci's thesis of *Complementing Confucianism and Replacing Buddhism* (*bu Ru yi Fo*). Xu Guangqi critiques Buddhism for its vagueness, lack of practicality, and potential to cause social instability, as well as its superstitious, immoral, and irrational cosmological beliefs (Zürcher 2001). He argues that these beliefs should be rejected and reinterpreted through the application of scientific methods, particularly astronomy, which requires mathematical calculations. In this regard, Euclid's works are perceived as a political tool that contributes to the enhancement of the Confucian state, as mathematics has practical applications in various fields including astronomy (i.e. the reform of the calendar), agriculture (i.e. water management and land division to increase productivity and tax collection), and military (i.e. fortification construction and weaponry improvement). Xu consistently evaluates Western works based on their usefulness for society (Hashimoto and Jami 2001, 263–278).

Therefore, the significance of these high-ranking officials in establishing and maintaining Confucianism should not be disregarded. They are also responsible for constructing and preserving Confucian temples, where their recruitment happens. Government schools serve as the locations for examinations and graduation ceremonies, thereby perpetuating the close relationship between the government and Confucian beliefs over time.The test to access to public office are based on the *Four Books*, the *Five Classics*, and historical records. The selection process is competitive and merit-based, centrally organized to prevent corruption and favouritism. Examiners from all provincial centres travel to Beijing to examine thousands of candidates, and candidates' answers are rendered anonymous before grading through a complex process that involves stamped signatures on the original answers and anonymous copies for assessment by the examiners. Ricci had a thorough understanding of the process, as evidenced by his statement: "if it is not possible to say of this realm that the philosophers are kings, at least one can say with truth that the kings are governed by philosophers."[5]

Confucianism was hence associated with prestige, prosperity, and influence and could serve as a suitable accompaniment for the missionaries whose objective was to render it as congruent as possible with Christianity. To propagate his teachings, Ricci opted not to employ a pulpit but instead presented himself to the Chinese as a philosopher and a sage, and his initial discourse was secular in the form of lectures rather than sermons. According to Gernet, this tactic –which he ascribes to discretion– explains why it took the Chinese a prolonged period to comprehend the true purpose of the missionaries' presence in their territory and the monotheistic essence of the Christian faith (Gernet 1982, 28). Ricci himself reflects upon this in one of his letters:

> I am firmly convinced that we will no longer establish a church, but instead a preaching house [*shuyuan* or *philosophical academy*], and we will celebrate Mass privately in another chapel, although for the time being, the room allocated for receiving visitors may also be utilized for this purpose as imparting teachings through conversational means is more effective than through sermons.[6]

Therefore, when France joined the other nations in the seventeenth century to steer their economic pursuits and their scientific and artistic curiosity towards China, they did so within an intellectual milieu that was well established by the Spaniards and Italians such as Ricci. It is not to say that France had shown no prior intellectual interest in Asia. A glance of the latest edition of Michel de Montaigne's *Essais* would be sufficient to dismiss such a notion. Nevertheless, all of Montaigne's sources were French or Latin translations of original Portuguese or Spanish texts that were disseminated into France and Germany through Italy (Lach 1977).

It happened that the development of France's presence in China was concomitant with the decline of Italian and Spanish influence in the region, and significant tensions with Rome. In the early seventeenth century, the Portuguese *patronage system*, which entailed the direct dependence of dioceses on the king rather than the Papacy, appeared to have reached its limits considering the demand for the establishment of new dioceses throughout the world (Mungello 1985, 24). Rome capitalized on this situation to initiate a project of centralizing evangelization, which took shape in the creation of the Congregation for the Propagation of the Faith (*Propaganda fide*) by Pope Gregory XV in 1622. Gregory XV was a reliable advocate of the Jesuits, from whom he had received his education (Lach 1998, 23). From then on, the Vicars Apostolic of the East would be solely dependent on the Papacy in both temporal and spiritual matters, and all missionaries departing for the East were required to take an oath of allegiance to the Pope, in which they accepted the hierarchical primacy of the papal vicars and acknowledged the superiority of papal authority over all others. These provisions were not to the taste of Louis XIV and

---

[5] Ricci in Laven 2011, 319. Translation mine.
[6] Ricci in Gernet 1982, 27–28. Translation mine.

his ministers nor were they well-received by the Jesuits, which gave rise to considerable tensions between them and the Vicars Apostolic in China (Pinot 1932, 46 ff).

It was in this controversial context that the European version of Zhu Xi's *Four Books* (*Si Shu*) was translated with an extensive commentary known as *Confucius Sinarum Philosophus* (1687).[7] The objective of the Jesuits in this translation was to address their critics, not only their Catholic contemporaries, but also Jansenists such as Arnauld and Huguenots like Jurieu. Maybe they also aimed to address Bayle, who at the time was already defending the possibility of an atheist society in his own work. This text is a compilation of the work of 17 Jesuits. The only signatories, however, were Ignatius da Costa (1603–1666), Prospero Intorcetta (1625–1696), and Philippe Couplet (1622–1693). Couplet, who was Belgian, had considerable evangelical activity in provinces such as *Jiangxi*, *Zhejiang*, and *Jiangnan*, but he also held the post of procurator in Europe for years. To find money and support for Jesuit missionary work in China, he travelled throughout Europe to build up a network of influential scholars and public figures, including Louis XIV's confessor François de la Chaise and minister Jean-Baptiste Colbert, and the Marquis de Louvois, who replaced Colbert as director of the French Academy of Sciences. Following their diplomatic efforts in France, a party of five eminent Jesuits was sent to China with the intention not only of strengthening the religious mission but also French influence there. These Jesuits were Joachim Bouvet, Jean-François Gerbillon, Louis le Comte, Guy Tachard, and Claude de Visdelou. This French sponsorship explains why Couplet dedicated the *Confucius Sinarum Philosophus* to Louis XIV. Bayle echoed Couplet's efforts in April 1686:

Approximately 18 months ago, when Chinese ambassadors were present in France, the Duke of Maine [Louis Auguste de Bourbon] informed the King that the Chinese had books on the history of their country dating back nearly 3000 years, that they were pioneers in the field of science and the arts; and that our understanding of these matters was not yet well-informed and, and that it was the duty of a prince such as the King and to import these books from China and to employ people for their translation. Then His Majesty gave orders at once for this project to be undertaken, and I am informed that a shipment of 300 volumes of Chinese books on civil history, natural history, mathematics, and other intriguing treatises has recently arrived in Paris. In addition, two translators have also arrived, one of whom is a Jesuit who has spent 30 years in China and the other is a Chinese individual from the recent embassy who is proficient in Latin, Italian, Portuguese, etc. These people will work diligently to translate the most valuable of these books, which will be published as soon as they are ready for distribution. Furthermore, as it is known that the Jesuits have good relations with the King of China, eight young men have been sent to his kingdom at the expense of the [French] King to learn the Chinese language and to teach the Chinese in both our language and Latin, with the aim of

---

[7] A recent English edition is available: *Confucius Sinarum Philosophus (1687). The First Translation of the Confucian Classics*. Thierry Meynard (ed.). Rome: Monumenta Historica Societatis Iesu, 2011. The original Latin edition is available online: https://play.google.com/books/reader?id=BerpG7rz0_YC&pg=GBS.PP4&hl=es

bringing them back to France to continue the translation work. Others will be brought in to teach us their mechanical arts. Another letter states that Father Couplet has returned from Rome, where he trained a Chinese Jesuits to follow in his footsteps, and it is anticipated that they will translate the complete works of Confucius (NRL: OD I, 537).[8]

The *Confucius Sinarum Philosophus* follows the categorization of *Chu Hsi*, who in the twelfth century grouped the texts *Ta hsüeh*, *Chung yung*, *Lun yü*, and *Mencius* as the *Four Books*. These texts, together with the *Five Classics* –commentaries on these books– constituted the *corpus* from which Chinese scholars were examined for becoming part of the government: the *I Ching*, the Book of Mutations (a divination manual); the *Shu Ching*, the Book of History (legislation); the *Shi Ching*, the Book of Poetry (an anthology of ancient poems); the *Chunqiu*, Spring and Autumn Annals (annals of the Confucian era); and the *Li Ching*, the Book of Rites (systems of government, philosophy, and ceremonies). However, the Jesuits do not accept the *I Ching* (the divination manual) as one of the classics since they interpret its provenance as non-Confucian. Perhaps the real reason was that the atheistic and materialistic philosophy of *I Ching* was in direct contradiction with the Christian teachings they wanted to spread there (Mungello 1985, 263).

Regardless, the Jesuits only translated the *Four Books*, despite their knowledge of the entire *corpus* of Chinese classics, in an attempt to separate the texts from their later commentators, to whom they attributed a certain leaning towards Buddhism or Taoism in their interpretations. Confucius is presented, once again, as a philosopher promoting spiritual and moral enhancement, but his religious significance—the veneration of his spiritual tablet—is muted. Similarly, translations that might approximate a divinized view of the earth are rejected in favour of an interpretation in line with creationist belief. Any metaphysical reference to the Confucian moral virtues—humanity *(Jen)*, justice *(I)*, ritual *(Li)*, and wisdom *(Chih)*—is conveniently ignored in favour of a civic interpretation. At other times, they force the

---

[8] "Il y a environ dix-huit mois que des Ambassadeurs de la Chine étant ici, Monsieur le Duc du Maine prit occasion de dire au Roi, que ces Peuples avoient des Livres de l'Histoire du pays depuis près de trois mille ans; qu'ils avoient connu les Sciences et les Arts tout des premiers; qu'on n'étoit pas encore bien informé du détail de tout cela, et qu'il n'appartenoit qu'à un Prince comme le Roi de faire venir de ces Livres de la Chine, et des gens pour les traduire. Sa Majesté donna aussi-tôt ses ordres pour ce projet, et l'on m'assûre qu'il est arrivé ces jours-ci à Paris jusqu'à trois cens volumes de Livres Chinois, tant d'Histoire Civile du païs que d'Histoire Naturelle, de Mathématique et d'autres Traitez curieux; qu'outre cela il est arrivé aussi deux Traducteurs dont l'un est un Jesuite qui a été trente ans dans ce Royaume; l'autre est un Chinois de la derniere Ambassade, qui sait le Latin, l'Italien, le Portugais, etc. Que. ces gens vont s'appliquer incessamment à traduire les plus curieux de ces Livres, qu'on fera imprimer aussi-tôt qu'ils seront en état de paroître. Et comme on sait que les Jesuites sont agréables au Roi de la Chine, on en a envoyé huit jeunes en son Royaume, auxquels notre Roi paye pension, pour apprendre la Langue du païs, et pour instruire des Chinois spirituels dans notre Langue et dans la Latine, afin de les faire venir en France pour continuer ces Traductions. On en fera venir d'autres pour nous apprendre leurs Arts Méchaniques. Une autre Lettre porte, que le P. Couplet est revenu de Rome où il a fait son petit Chinois Jesuite comme lui, et qu'on espere qu'ils traduiront toutes les Oeuvres de Confucius" (NRL: OD I, 537). Translation mine.

meaning of some concepts to bring the Chinese closer to Christianity on the belief in saints, in a supreme divinity, or in the immortality of the soul (Mungello 1985, 268 ff). The impact of the *Confucius Sinarum Philosophus* in Europe was considerable. Between 1687–1688, it appeared in the *Journal des savans* (Paris), the *Biblioteque universelle et historique* (Amsterdam), and the *Acta eruditorum* (Leipzig). Bayle also refers to it in his *Réponse aux Questions d'un Provincial:*

> *The sect of the savants or erudite have gained significant recognition, despite not being the most prevalent in China. The savants originated in the 1070s during The Emperors of the Sung dynasty, who loved literature. However, it did not make significant progress until 1400 CE, when the emperor* Yum lo *selected forty-two of the most talented doctors and commissioned them to compile a body of doctrine based on the classical writings of the ancients, with a particular focus on the philosophers* Confucius *and* Mencius (RQP: OD III, 926).[9]

However, we know that the purpose of the missionaries was not to explore new cultures and satisfy the curiosity of scholars but to convert new neophytes. It is not surprising, therefore, that Bayle was selective in his selection and interpretation of the data they suggested, or in his choice of their arguments. This explains why, as Juliette Charnley points out, the philosopher of Rotterdam tended to ignore or criticize everything that was not relevant to his philosophical-political purposes and to focus on issues that allowed him to reinforce his owns ideas, for instance civic tolerance or moral atheism (Charnley 1990, 181). However, Bayle also provides an account of many other details of Chinese culture that are not related to those issues. In his texts, he speaks of tea and its benefits (NRL: OD I, 233 and 286); talks about the use of opium (NRL: OD I, 695); summarizes its remedies for gout (NRL: OD I, 234); describes the immense trees that cover up to 200 sheep (NRL: OD I, 694); informs of parents who sell their children and summarizes the custom of infanticide (RQP: OD III-II, 710–711); speaks of Chinese pronunciation (NRL: OD I, 164); and discusses the creation of the Chinese wall to stop Tartar raids (NRL: OD I, 231 ff). He also reports that the Chinese were familiar with gunpowder long before it was known in Europe (NRL, OD I, 214). Chinese medicine also has its place in the work of the philosopher of Rotterdam, who notes that there were doctors in China 2697 years before the Common Era (NRL: OD I, 638). He also points out that, when applied to women, medicine encounters the same problem that religious proselytizing faces: access to women is restricted to their husbands and relatives. Indeed, missionaries meet enormous difficulties in converting Chinese women since they are not allowed to be in the company of or converse with men other than their husbands (RQP: OD III, 1039). For that reason, when a woman is ill, the doctor must guess her condition by taking her pulse, as they are forbidden to inspect her body.

---

[9] *"La secte des sçavans ou lettrez est devenuë la plus celebre, quoy qu'elle ne soit pas la plus commune dans la Chine. Elle commença vers l'an mil soixante et dix, sous les Empereurs de la race de* Sum, *qui aimoient les lettres; mais elle fit peu de progrès jusqu'à l'an mil quatre cens, que l'Empereur* Y um lo *choisit quarante-deux Docteurs des plus habiles, auxquels il ordonna de faire un corps de doctrine tirée des livres Classiques des Anciens et particulierement des Philosophes* Confucius *et* Mencius" (RQP: OD III, 926). Original text in italics. Translation mine.

Bayle ironizes by saying that this is a strange custom which assumes that, instead of pity, a sick woman could provoke an erotic reaction in her doctor. For him, Chinese medicine is much more regressive than European medicine, which, in addition to not relying on pulse measurement, allows the dissection and study of corpses, which is forbidden in China (NRL: OD I, 638). Bayle also echoes the problems the Jesuits have in explaining the dogma of transubstantiation in those lands (NRL: OD I, 712) or their refusal of the political use of Hell that Christians implement as a means of controlling people (RQP: OD III, 963). He comments on their belief in reincarnation in animals (DHC, Hali-Beigh, C) and uses their example to compare their beliefs about the soul with those of Lucretius (DHC, Lucrèce, R). Of course, judicial astrology is another subject in which the Chinese provide a paradigmatic example that is more than familiar to Bayle's readers (NRL: OD I, 707; PD: OD III, 20 ff.). But, above all, Bayle considers China to be the wisest and most ingenious country of the East: Are you aware that Spinoza's atheism is the doctrine of several sects that are prevalent in Asia, and that among the Chinese, the most learned and ingenious nation in the East, there is a sect of atheists that is composed of the majority of the literati or philosophers? (CPD: OD III, 210).

However, Bayle's references to China and its customs do not only come from the Jesuits. The philosopher of Rotterdam also cites the Italian Pietro Della Valle (1586–1652), whose travels were published in *Viaggi di Pietro Della Valle il pellegrino* (1650–53), and French royal librarian Melchisedec de Thévenot (1620–1692), whose collection of travel literature contained several volumes on China. He also draws on *Description du Royaume de Siam* by French diplomat Simon de la Loubère (1642–1729), which is also cited in his *Dictionnaire* (DHC, Sommona-Codom; Ruggeri, n.32; Spinoza, n. 147,148). Bayle also collects data from Le Chevalier de Chaumont (1640–1710), the first French ambassador to Siam. Later, he adds to his sources the Augustinian Juan Gonzalez de Mendoza, author of *Historia de las cosas más notables, ritos y costumbres del gran reyno de la China* (Rome, 1585), a work that went through more than 38 editions. Mendoza never travelled to China (Sola 2018), although Bayle refers to him in his *Dictionnaire* as the ambassador of the King of Spain (DHC, Mendozza). Mendoza's work was recognized as the beginning of modern sinology, as it incorporated the most recent information from the first Spanish embassy to China in 1575 and did not rely upon Marco Polo's accounts from the Middle Ages (Charnley 1990).

Given these facts, it is difficult to doubt Bayle's extensive knowledge of literature on the East, although it must be noted that it is not always clear whether he read these works directly, as much of the information he possesses is derived from the Jesuit Catalogue of 1686: *Catalogus Patrum Societatis Jesu qui post obitum S. Francisci Xaverii ab anno 1581 usque ad 1681 in imperio Sinarum Jesu Christi fidem propagarunt* (NRL: OD I, 662 ff). However, this does not suggest that Bayle's understanding of Oriental philosophy was superficial, as Thijs Weststeijn notes (2007). Rather, Bayle's knowledge of China is demonstrated through its constant presence in his work, which is comparable to the interest shown by his contemporaries in that far-off land.

4.3 Bayle's Rejection of Natural Religion

The discussion on natural religion was a question that theologians asked themselves when they encountered non-Christian societies. The dilemma was whether these communities should be regarded as being saved by God, despite their ignorance of Christianity, or if they should be condemned due to their lack of knowledge. This question arose with the encounters of Spanish Catholic missionaries in the Americas and became even more delicate when it came to the peoples of the East. If the answer was positive, it would reinforce the idea of a common natural religion. This was the position adopted by Jesuit missionaries in the East, which implied adopting a more flexible idea of dogmatics and ceremonial organization, or even an appeal to a pre-biblical past in which the East was a privileged place of Sacred History as Joachin Bouvet (1662–1732) and the *figurists* argued.[10] On the other hand, if one chose to disassociate these groups from Sacred History, one would have to explain the reason for their prosperity and the moral progress shown by their inhabitants. After all, Christian theology has always linked public prosperity and moral evolution to knowledge of the true religion—their own–. That's why China posed a challenge for those who defend the link between religious truth and social peace; in other words, those who believe in the necessity of the spiritual mediation of the church and clergy in moral progress. Nevertheless, advocating for a universal natural religion was not an easier solution to the problem, as it meant downplaying the importance of ceremonies. In other words, that meant rejecting the link between religious belief and a particular dogmatic system and thus the very need to belong to a particular church to be saved. And as we know, this is the essential core of the thesis of *compelle intrare*, the heart of religious coercion, which will be a constant critical element in Bayle's work on tolerance.

Indeed, before knowing in detail what the Jesuits had to say about China, Bayle's work addressed the role of rites and ceremonies in terms of their good or bad alignment with the passions. Thus, in 1683, when he wrote the *Pensées diverses*, he observed that religious ceremonies provided a remarkable resilience to change among followers of any faith, even when it involved rites that were difficult to observe. According to Bayle, this is because ceremonies are perfectly suited to the passions.[11] It is the rites, rather than the knowledge of the moral principles that every

---

[10] The so-called Figurists maintain that they were personages in the Bible that may possess multiple interpretations. The Jansenist Abbé d'Etémare was the first to put forth this idea, but it was Bouvet who spread it most effectively through his extensive correspondence with scholars across Europe, including Leibniz. Figurists continued the legacy of the *Prisca Theologia* or Hermeticism tradition, which held that certain elements of paganism may contain residues of the truth of religion. Supporters of this perspective include Lactantius (240–320), Hermes Trismegistus, Pico della Mirandola, Giordano Bruno, and the Cambridge Neoplatonists.

[11] "Ou bien disons, que s'ils observent régulierement plusieurs cultes pénibles et incommodes, c'est parce qu'ils veulent racheter par là leurs péchez d'habitude, et accorder leur conscience avec leurs passions favorites; ce qui montre toujours, que la corruption de leur volonté est la principale raison qui les détermine" (PD: OD III, 89).

religion claims to impart, that bind people together in the same community of faith. That is demonstrated by circumcision among the Jews or incestuous marriages among the Persians (PD: OD III, 88). And it is these diverse ritual practices that set men against each other, showing that they are not guided by universal moral principles but by passions.[12] As a result, the presence of this diversity of rituals highlights the prioritization of ceremonial observances over the ethical core of religions. It is for this reason that fear of God has never constrained criminal behaviour. Therefore, the connection between political obedience and the adoption of specific ceremonies –whether it be a Christian confessional state or a mandatory civil religion– does not lead to atheism but rather to fanaticism:

> It has been recognized in all times that religion was one of the bonds of society and that the subjects were never kept in a state of obedience better than when one could have the minister of the gods intervene; and that one was never able to encourage the people with greater success in the defense of the fatherland than by attaching their heart to certain devotions practiced in certain temples with pompous ceremonies, under the protection, a thousand times experienced, of certain divinities, and by making them believe that the enemies who wished to profane these holy places were threatened with a terrible punishment according to the presages of the victims. To set all these springs in motion, there had to be, not only a religion authorized by the magistrate, but also subjects informed by fear, veneration, and respect for all the exercises of this religion. This is why politics wished to have managed carefully all that would be suited to foment the zeal for religion in men's minds and to inspire in them a profound respect for its smallest ceremonies. Judge, Monsieur, whether in accordance with this there was reason to fear that the people would fall into atheism. (PD: OD III, 73/ 2000, 139).

In this sense, both the Christian confessional State (State religion) and systems governed by a civil religion have the potential to exhibit both tolerance and intolerance as pagan religion revealed. The distinction between tolerant and intolerant systems is not just the absence of an obligatory doctrine of faith, but also the lack of a compulsory observance of rituals. It is crucial that both these requirements are fulfilled. If this is not the case, it could be said that Emperor Augustus, Calvin and Rousseau would be on the same level since they all support a form of politics in which religious ceremonies are a crucial piece of political allegiance. This system creates the possibility for religious-based disobedience as any violation of the cult that is required by the State is considered a disruption of public order and civil peace (García-Alonso 2019).[13] Likewise, when a religion becomes so closely tied to its rituals, it creates the requirement for exegetes and intermediaries between God and humanity. In ancient Rome, the function of this new clerical caste was seen as the

---

[12] "généralement parlant (car j'excepte toûjours ceux qui sont conduits par l'Esprit de Dieu) la foi que l'on a pour une Religion n'est pas la regle de la conduite de l'homme, si ce n'est qu'elle est souvent fort propre à exciter dans son ame de la colere contre ceux qui sont de différent sentiment, de la crainte quand on se croit menacé de quelque péril, et quelques autres passions semblables, et sur tout un je ne sai quel zele pour la pratique des cérémonies extérieures" (PD: OD III, 92).

[13] We use here the term *State* in a broad and anachronistic sense, as the *modern State* was still in development in the seventeenth century. Moreover, we do not distinguish between different systems of government, even though political tolerance may take different forms as an oligarchic republic like the Netherlands, a monarchy like England or France, or an Empire like China.

only way to avoid the problems facing the state (PD: OD III, 73). The paradigmatic example of such perversion, as could not be otherwise, is the persecution of heretics, which Bayle addresses at length in all his works. For all these reasons, Bayle held that it is not the knowledge of religious principles that moves humanity, but only the passions (PD: OD III, 89). And this rule of passions is ineradicable and becomes effective even if it means sacrificing the principles we claim to believe in (PD: OD III, 91). It is a classical principle that Bayle borrows both from Ovid—*video meliora proboque, deteriora sequor*—and from Paul in his Epistle to the Romans (Romans 7:15) (DHC, Ovide H; Hélène Y). Therefore, let us consider that what Bayle describes as the rule of passions is a universal epistemic or psychological principle, as it affects all individuals regardless of their religion or social class. Accepting this fact is the only way to understand why many Christians act immorally despite having knowledge of revelation, experiencing miracles, receiving guidance from excellent preachers or spiritual directors of conscience, and possessing devotionals (PD: OD III, 87). As this psychological principle is universal, it also applies to idolaters or atheists. It is the desire for pleasure and the avoidance of pain that motivates all human action, not religious values (PD: OD III, 106/Bayle 2000, 139).

For since experience shows us that those who believe in a paradise and a hell are capable of committing every sort of crime, it is evident that the inclination to act badly does not stem from the fact that one is ignorant of the existence of God and it is not corrected by the knowledge one acquires of a God who punishes and rewards. It follows manifestly from this that the inclination to act badly is not found in a soul destitute of the knowledge of God any more than in a soul that knows God; and that a soul destitute of the knowledge of God is no freer of the brake that represses the malignity of the heart than is a soul that has this knowledge. It follows from this in addition that the inclination to act badly comes from the ground of man's nature and that it is strengthened by the passions, which, coming from the temperament as their source, are subsequently modified in many ways according to the various accidents of life. Finally, it follows from this that the inclination to pity, to sobriety, to good-natured conduct, and so forth, does not stem from the fact that one knows there to be a God (for otherwise it would be necessary to say that there has never been a cruel and drunken pagan) but from a certain disposition of the temperament, fortified by education, by personal interest, by the desire to be praised, by the instinct of reason, or by similar motives that are met with in an atheist as well as in other men (PD: OD III, 94/Bayle 2000, 180).

This quote reveals the three key components of Bayle's anthropology. The passions are the universal psychological factors that are expressed differently in everyone, shaped by their temperament and the customs of the society in which they were raised. Each person has a unique personality that influences how they experience the universal passions, which in turn determines their character and personal preferences. However, temperament and customs are not the only factors that structure the expression of the passions. In that sense, religion cannot be the source of moral behaviour because individuals do not act in accordance with their beliefs, but rather according to their temperament, preferences, and education. Bayle provides numerous examples of this disconnect between belief and practice, arguing that without this concrete component, the classical idea that piety or belief in gods and religion is the foundation of morality and social harmony is meaningless. On the contrary, religion may even exacerbate the passions, as demonstrated by the violence that some religions inflict upon others.

Starting in 1684, when he began writing his *Nouvelles de la République des lettres*, Bayle's views became even more radical. In this journal, he drew upon the works of those who provided him with information on non-European civilizations, as previously mentioned. This included accounts from both colonization and religious missions, which demonstrated that atheism is an ethnographic reality that can be found among the Chinese, Africans, and inhabitants of the Mariana Islands in Oceania, even in societies that are not highly developed. As demonstrated by John Christian Laursen in this volume, Bayle's familiarity with the religious beliefs of the indigenous peoples of the Americas was limited. The evidence they supplied augmented his overall thesis regarding atheism, yet it did not seem to noticeably modify his overarching argument. For instance, Bayle posited that the Cafres and Iroquois, who he understood as atheists, were no more uncivilized or uncultured than many other communities discovered recently in Africa and the Americas. While variations in the industry, ferocity, and morality among barbarian peoples may exist, Bayle challenged the notion that an individual who venerates a rock, wood, or stream necessarily possesses a greater degree of intelligence and reasoning than someone who does not worship anything, as pointed out by Laursen in this volume.

As demonstrated by the prevalence of negative atheism among the American and African populations, it can be argued that religion can be considered as a contingent historical outcome, and it is not a universal phenomenon. It can therefore be inferred that it has supplanted the atheism that characterized the state of nature in which mankind existed (Mori 2004, 383). Therefore, it can be concluded that religion is not an indispensable component of societies. We agree with Mori and Gros (2014) that the notion of an atheist society enables him to stress that religion is not required to explain the social bond that gives rise to the formation of political communities. Only civil legislation is necessary. It is impossible not to discern echoes of Hobbes in this paragraph:

> A state in which one constantly fears being plundered or murdered and must always be prepared to either defeat or defend against enemies is too violent for one not to desire to put an end to it. The desire to escape such a state is one of the reasons for the formation of human societies. It can therefore be argued that the peoples who lived without any form of government, divided into independent families, did not disturb their neighbours. If some of these people were eventually forced to form a political body to live more securely, this was a sufficient reason for their coming together. If they believed in gods, they continued to worship them; if they did not believe in gods, they continued to worship none, and therefore religion was neither the motive nor the foundation for their confederations. The same interest that led to the formation of these political bodies has continued to maintain them. It is

therefore not essential for the maintenance of political bodies that they have a religion (CPD: OD III, 353).[14]

Morality does not require a religious root. In evident contrast to America or Africa, China has a high level of complexity and development. Thus, it is difficult to support the idea that morality is tied to religiosity as theologians do, when we have knowledge of a people who have lived for centuries without knowing Christianity, yet they have impressive cultural and moral achievements. Besides, the atheism of the Chinese is not merely a negative absence of belief, like that of the Americans or Africans,[15] but rather a conscious rejection of all *ontotheology*:

> The numerous reasons that have been cited as evidence of the existence of atheistic peoples in the New World, and the credibility of the authors who support these claims, are a source of considerable distress for theologians who follow the ideas of Mr. Bernard; however, nothing is more likely to cause them concern than the reports about the atheism of the philosophers of China. This is not a simple, negative atheism like that of the American savages; it is a positive atheism, as these philosophers have compared the idea of the existence of God with the opposed system (RQP: OD III, 925).[16]

Chine has a philosophical atheism (positive atheism), not just a cultural (negative) one (RQP: OD III, 926). The atheism of the Chinese is not a mythological atheism like that of the Americans (CPD: OD III, 352), but rather a real atheism that is supported by the accounts of missionaries who travelled to China (RQP: OD III, 927; DHC, Maldonat, L) and reveals to Europe that the main objective of this atheism is

---

[14] "Un état où l'on craint d'être pillé et assassiné à toute heure, et où l'on se doit toûjours tenir prêt soit à primer, soit à repousser les ennemis, est trop violent pour qu'on ne souhaite pas d'y mettre une fin. La nécessité de s'en délivrer passe pour l'une des causes des Sociétez humaines. Il faut donc dire que les peuples qui ont vécu sans aucune forme de gouvernement, et divisez en familles indépendantes, ne troubloient point le repos de leurs voisins. Si quelques-uns ont été enfin contraints de former un corps de société pour vivre plus sûrement, voilà une cause sufisante de leur réunion. S'ils croïoient des Dieux, ils continuoient à les servir; s'ils n'en croïoient pas, ils continuoient à n'en servir point, et ainsi la Religion n'a été ni le motif, ni la base de leurs confédérations. Or le même intérêt qui les a formées au commencement, a continué de les maintenir. Il n'est donc pas d'une absoluë nécessité pour le maintien des corps politiques qu'ils aïent une Religion" (CPD: OD III, 353). Translation mine.

[15] "Il y a eu des Nations qui ont subsisté sans loix, sans Magistrats, sans aucune forme de gouvernement. Les Aborigines en Italie, les Gétules et les Libyens en Afrique se sont maintenus ainsi pendant plusieurs siécles […] Cette maniere de vivre est fort mal propre à polir les mœurs, et laisse l'esprit dans une stupidité sauvage; mais il n'est pas question de cela, […] Quoi qu'il en soit, voilà des peuples qui se sont multipliez et conservez sans vivre en Société. Il est donc faux que la vie sociale soit absolument nécessaire à la conservation du genre humain. […] ces peuples qui n'avoient aucune forme de gouvernement, ni aucune loi, il y en avoit d'Athées" (CPD: OD III, 352).

[16] "Le grand nombre de relations qui assûrent que l'on a trouvé des Peuples athées dans le Nouveau Monde, et le mérite des Auteurs qui ajoûtent foi à cela, font beaucoup de peine aux Théologiens dont Mr. Bernard adopte les hypotheses; mais rien n'est plus propre à les chagriner que les témoignages qui concernent l'Athéisme des Philosophes de la Chine. Ce n'est pas un simple Athéisme négatif, comme celui des Sauvages de l'Amérique; c'est un Athéisme positif: car ces Philosophes ont comparé ensemble le système de l'existence de Dieu, et le système opposé" (RQP: OD III, 925). Translation mine.

to maintain social order.¹⁷ Even more, Bayle's opposition to the idea of a universal natural religion was not temporary, as he maintained this view in his later works, such as *Réponse aux questions d'un Provincial* (1703–1707):

> If religion were an inherent characteristic of humanity, it would not be necessary to travel or rely on reports of explorers to confirm that all people around the world acknowledge and have always recognized the existence of a deity. To establish this universal truth, we need only consider our understanding of the nature of humanity. This truth would be manifest, just as it is self-evident that all humans are rational beings. However, because religion is an accidental attribute of our species, as demonstrated by the fact that one can still accurately describe a person's essential characteristics even if they do not possess religious beliefs, it cannot be assumed without evidence that all people across the globe agree on the existence of a deity (RQP: OD II, 693).¹⁸

Therefore, contrary to what the Jesuits argue, there is no natural religion and only the action of education and custom can explain the reality that there are peoples who have no religious beliefs at all.¹⁹ For this reason, we disagree with Simon Kow when he argues that Bayle maintained his skepticism in the debate on Confucianism –without deciding whether it is a natural religion or an atheistic system (2017, 64)–. On the contrary, we believe that Bayle's originality relies in the idea that China represents the certainty of a civilization that had progressed for centuries in the absence of any religion whatsoever. Bayle repeatedly argued that principles do not guide

---

¹⁷ "Le Dieu Fo est la principale idole de la Chine. [...] Bonzes une secte particuliere d'Athées, fondé sur ces dernieres paroles de leur maistre. [...] *La fin que le sage se propose est uniquement le bien public. Pour y travailler avec succès, il doit s'appliquer à détruire ses passions, sans quoi il luy est impossible d'acquerir la Sainteté, qui seule le met en estat de gouverner le monde, et de rendre les hommes heureux. Or cette Sainteté consiste dans une parfaite conformité de ses pensées, de ses paroles et de ses actions avec la droite raison. Ce n'est pas que les passions soient mauvaises, quand on en sçait faire un bon usage; mais comme elles troublent presque toûjours la tranquillité de l'esprit, il faut en retrancher la trop grande vivacité, et faire ensorte qu'elles ne soient plus des emportemens outrez de la cupidité, mais de justes sentimens de la nature.* Prenez la peine de considérer cette morale, et jugez après cela si ceux qui nient l'existence de la Divinité sont privez nécessairement des notions par où l'on discerne la vertu d'avec le vice" (CPD: OD III, 397). Italics refers to Le Gobien quoted literally by Bayle.

¹⁸ "Si la religion étoit un attribut essentiel à l'espece humaine, on n'auroit besoin ni de voïages, ni de relations de voïages, pour pouvoir dire véritablement que tous les peuples du monde reconnoissent la Divinité, et l'ont toûjours reconnuë, et la reconnoîtront toûjours. Il ne faudroit pour être bien assûré de cette proposition universelle, que considérer l'idée que l'on a de l'homme. On y verroit clairement cette vérité, sans s'être servi de l'induction, comme l'on y voit sans cette voie que tous les hommes sont des animaux raisonnables. Mais parce que la religion est accidentelle à notre espece, comme il paroît de ce que si l'on supose qu'un homme est privé de religion, on ne laisse pas de conoître que l'essence, ou que la définition de l'homme lui convient encore totalement; on ne peut être assûré, sans se servir de l'induction, que tous les peuples de la terre donnent leur consentement à l'existence divine" (*RQP*: OD II, 693). Translation mine.

¹⁹ "Cela fait penser qu'ils la reconoissent par un instinct de la nature, et que c'est un caractere indélébile avec lequel nous naissons tous, ou que pour le moins l'existence de la Divinité est si clairement marquée dans les ouvrages de la Nature, qu'un chacun s'en apercevroit, quoi que personne ne lui en parlât. On ne peut point décider ainsi, lors qu'on fait qu'il y a des peuples assez stupides pour ignorer qu'il y ait un Dieu. On se croit alors obligé de dire que la Religion est une chose qu'il faut apprendre, et que l'homme ne sauroit pas s'il n'y étoit élevé" (RQP: OD III, 696).

action and that the majority's conduct is governed by the law or by the promise of honour and glory that comes with conformity to social norms. Therefore, Bayle asserts that concrete, historical, positive human law is required for maintaining civil peace, rather than piety, as was defended by Cicero (PD: OD III, 84). It must be acknowledged that the social bond must stem from a source prior to religion, as it has been demonstrated that not everyone is religious (atheists) nor does everyone follow the true religion (pagans, heretics) (PD: OD III, 84). Based on these premises, it is only logical to accept that human law is the only means of controlling immoral or uncivil behaviour. Its function is focused on restraining behaviours, rather than promoting the moral development of individuals. Hence, Bayle maintains that if religious teachings were the sole guiding force in society, practices such as incest and adultery would likely become widespread even in Christian communities (PD: OD III, 88). Therefore, human law is a more effective source of obedience than fear of the gods, even for Christians, as they fear the sovereign more than divine punishments, as the latter are more immediate and close at hand.[20] Thus, the principles that serve to suppress social and moral chaos are justice and honour, which govern societies that lack the knowledge of the true religion (CPD: OD III, 372).

Consequently, coercion is a fundamental aspect of any human community and even in Christian societies, civil law is essential for preserving social peace. However, another law is also necessary for the establishment of political states, one that is internal and not imposed through coercion, but rather through incentives such as the desire for honour and glory. This desire comes not from religion, but rather from human nature (PD: OD III, 115). Adherence to this internal rule can be attained through various means such as persuasion—through religious teachings or exchanging ideas—or through social pressure. For instance, in repaying a loan, the borrower's sense of pride or shame for breaking their promise may serve as stronger motivators than greed. In a similar manner, Roman women esteemed modesty and chastity for the purpose of societal recognition rather than religious justifications, since their gods were widely recognized for their promiscuity. Thus, religion may reinforce a behaviour, but it is not its source.[21] This is demonstrated by the presence of atheists, whose yearning for posthumous glory cannot be accounted for by their belief in the immortality of the soul (PD: OD III, 110; RQP: OD III, 966). Social pressure is an effective means of controlling misconduct without violence. Politically speaking, the drive for honour and glory serves as a more effective warning against social conflict than religious belief, as maintaining political stability is crucial for

---

[20] "D'où vient donc que la crainte des hommes est plus active sur eux que celle de Dieu? C'est que celle-ci ne considére son objet qu'en éloignement et avec les yeux de la foi, et que l'autre se raporte à un mal visible, certain et prochain" (CPD, OD III, 386).

[21] "Ce qui n'empêche pas qu'on ne puisse dire, que la Religion se mêle souvent dans ce ressort, et qu'elle lui donne de grandes forces pour les choses, où le tempérament nous incline: par exemple, un homme bilieux est bien-tôt armé de zêle contre ceux qui ne sont pas de sa Secte. C'est la foi, dit-on, qui est cause de cela. Dites plûtôt, que c'est l'envie naturelle, et le plaisir que nous avons tous de surpasser nos rivaux, et de nous venger de ceux qui condamnent notre conduite" (PD: OD III, 11).

the success of this ambition (CPD: OD III, 358). Persuasion is another means of controlling, modifying, or arousing passions with the goal of shaping individual convictions. This is the essence of the imaginary dialogue in which Bayle presents the Chinese as impartial arbiters who are willing to make a rational choice between different theological interpretations regarding the Eucharist:

> *Dear Christian Missionaries,* (they would address the controversialists), *coming from such distant lands to inform us that you disagree with one another, we cannot attend to your disputes for as long as it takes. And as you have mentioned Socinians, Independents, Episcopalians, as well as other Sects, it would be fair for us to hear from them too. Please request that they send their representatives here; they may enlighten us. In the meantime, we are not afraid of you, you will not convince any Chinese, unless you rely just on reason and on the condition that the emperor prohibits all of his subjects from adopting Christianity* (CP: OD II, 522).[22]

Thus, the means of controlling the impact of human emotions on society lies in the use of coercion and social pressure, rather than religion. For this reason, Weber's concept of the *disenchantment of the world* (*Entzauberung der Welt*), which Gros uses to interpret Bayle (2014), does not imply a secularization of society but rather a political model that does not require the removal of religion from the personal beliefs of its members, but only from the state. For that reason, this model does not require the privatization of faith but just a non-confessional state. Furthermore, in China, it is the atheist philosophers who hold positions of power and influence in the government. What more proof than the existence of China is needed to argue that politics does not require any religious legitimacy to flourish, and religion and politics do not have to go hand in hand?

4.4 Baylean Political Atheism

Bayle's rejection of the existence of a universal religion is not merely a theoretical debate that follows the format of a comparison between pagans and Christians. Indeed, from the moment he obtained ethnographic data on China from missionaries, Bayle discourse extended beyond a hermeneutic analysis of texts written by historians, jurists, or philosophers. We agree with Pinot's claim that debates about China were not confined to the private schoolwork of philosophers, but rather took place in the public sphere (1932, 428). The discussion on China was highly influential and numerous books on the subject were published and widely distributed

---

[22] "*Messieurs les Convertisseurs Chretiens,* (diroient-ils aux Parties contestanstes) *qui venez de si loin pour nous aprendre que vous n'êtes pas d'accord entre vous, nous ne saurions vaquer à vos disputes tout le tems qu'il seroit nécessaire; et puis que vous avez fait mention de Sociniens, d'Indépendans, d'Episcopaux, comme d'autres Sectes, il seroit juste que nous les entendissions aussi; mandez-leur qu'ils envoïent ici leurs Députez; peut-être nous fourniront-ils des lumieres. En attendant nous ne vous craignons gueres, vous ne gagnerez aucun Chinois, pourvû que vous ne vous serviez que de la Raison, et pourvû que l'Empereur défende à tous ses Sujets d'embrasser le Christianisme*"(CP: OD II, 522). Italics in the original text. Translation mine.

throughout Europe, as noted by Thijs Weststeijn (2007). These manuscripts demonstrate that while Europe was experiencing the destabilizing effects of conflict between different Christian denominations, China was enjoying a period of stability. And Bayle wondered whether this stability was due to or despite the fact that China was governed by atheists.

As we have said, Bayle shares with the Jesuit missionaries the idea of Confucianism as a moral doctrine. Yet, despite Ricci's emphasis on sincerity being one of the foundational principles of Confucian ethics, he also acknowledged that every Confucian civic ceremony was intrinsically tied to the effective governance of the kingdom. These remarks may have prompted Simon de La Loubère to question the piety or faith of Chinese rulers and to suggest that they only adopted the public ceremonies or rituals of the dominant religion for political reasons, rather than out of any genuine conviction. La Loubère's thesis is adopted by Bayle in his *Dictionnaire* (DHC, Ruggeri D), and is also supported by Vossius, whom Bayle cites extensively. However, it is important to note that Bayle's portrayal of Confucianism may not accurately represent actual Chinese practices. Actually, devotion to Confucius was officialised in a ceremony in 1530, where his wooden tablet –believed to embody his spirit– was worshipped. The Chinese Ministry of Rites subsequently named him Master K'ung the Most Holy Teacher of Antiquity (*chih sheng hsien shih K'ung Tzu*). This form of aristocratic civic worship was well-known throughout Chinese society and was the type of Confucianism that the first Jesuit missionaries encountered upon their arrival in China. One of the core beliefs of Confucianism was the observance of *hsiao* or filial piety, which entailed the duty to care for one's parents, provide them with a dignified funeral and burial, and show obedience and respect to them. Over time, the scope of filial piety expanded from just parents to lineage, society, and even to the dead, represented by wooden tablets that were honoured by each family head with specific rituals. *Hsiao* was a fundamental virtue that maintained the stability of Chinese society and proved the harmony between heaven and earth. Furthermore, the cult of Confucius was integrated into government schools as early as 59 A.D., which partly explains its longevity despite the introduction of Buddhism and Taoism into Chinese society. As every government minister was educated in these schools and belonged to the elite who were familiar with Confucius's works, his ideas were a central aspect of their studies. Nevertheless, the Jesuits never accepted that Confucianism had a religious interpretation of Heaven that comprehended the universal order of morality, rituals, and the socio-political arrangement. That justified that the establishment of the calendar was exclusively an imperial power, demonstrating the intricate connection between time and space that was entirely under the control of the Emperor. The celebration of Confucian ceremonies was also the exclusive responsibility of the Emperor. Hence, Bayle's perception of Confucius is shaped by the sources he consulted.

Indeed, as Lorenzo Bianchi notes, the debates contained in *Pensées diverses* on whether atheism is worse than superstition draw from previous sources such as La Mothe le Vayer, and are therefore not entirely original (Bianchi 2011). It was La Mothe le Vayer who was the first philosopher to bring Confucius alongside the

Greeks and Romans into the discourse on the morals of pagans. And it was Gianni Paganini who illustrated the close connection between Bayle and La Mothe le Vayer, both in terms of the philosophical approach they used to argue with their opponents and the beliefs they upheld. They both reject the dogmatic truth embraced by theologians and highlight the dependence of our opinions on education and emotions (1980, 88). However, in our opinion, when it comes to Confucius, their views diverge. In *De la vertu des Payens* La Mothe le Vayer suggests that the Chinese could have achieved salvation simply by following moral principles that align with natural law. Unlike the Romans and Greeks, their monotheism is reflected in their belief in one King of Heaven, and their morals align with the natural principles of Christianity. Confucius, therefore, is seen not as a deity but as a philosopher revered for his morality. His impact at court was so significant that under his influence an empire in China was established, ruled by philosophers and reason, in which loyalty to the prince, love for the country, and the maintenance of civil peace were highlighted, all without the use of military force, as the Japanese did: "What greater happiness can one ever wish for than to see kings be philosophers or philosophers reign?" (Le Vayer 1642, 283). These are arguments that follow the theses of Matteo Ricci, whom Le Vayer cites as a direct source in his work. However, Le Vayer adds something else: Confucius may have received a special grace from God, not only for his virtuousness but also for his belief in this singular deity (1642, 291). Consequently, Confucius is portrayed not as a virtuous atheist, as he would be in Bayle's view, but as a monotheist who follows the universal natural law, which is the foundation of Christianity itself. Certainly, Le Vayer said that Confucius followed a natural religion that avoided any reference to original sin, which was considered scandalous in his time and was heavily criticised by his contemporaries, who saw in his works a revival of the Pelagianism condemned by Augustine, which held that man could save himself through his own means. Nonetheless, he assumed that Confucius was by no means an atheist. Yet, Bayle argues that Confucianism could be seen as equivalent to philosophical atheism.

Gianni Paganini highlights that Bayle recognizes several varieties of atheism, which he delineates in his writings, yet can be divided into two categories (2009). The first includes all those individuals who lack knowledge of God, either due to ignorance or because the concept is foreign to their cultural milieu, such as African atheists, primitive communities, or Chinese Confucian elites. The second group comprises both those who are familiar with the concept of God but reject the rationality of the notion –speculative atheists such as Spinoza or Strato–, and those who are unable to either affirm or deny the notion of God –sceptical atheism–.[23] Followers of the Foë sect do not believe in either Providence or the immortality of the soul

---

[23] According to Paganini, Bayle would fall within the latter group of sceptical atheists, who comprehend the concept of God but are not persuaded of its existence through natural theology nor do they find materialist ontology a sufficient system for explaining the world. This sceptical atheism, which foreshadows that of Hume, is forged within an anthropological framework where education, ignorance, and instinct are given equal standing to divine grace. What is critical are the effects of action rather than the principles that drive it (Paganini 2009, 406).

(RQP OD III, 926; DHC, Spinoza B). Supporters of Confucius believe in Providence and deny the existence of a deity, unlike Epicurus, who, according to Bayle, believes in a deity but denies Providence (DHC, Sommona-Codom, A). This opinion can also be found among Muslims (PD: OD III, 110–11). What is noteworthy is that despite their differing metaphysics or ontologies, all atheists, according to Bayle, such as Pomponius Atticus, Cassius, Epicurus, Lucretius, Diagoras, Pliny, and Panetius, are models of moral virtue (CPD: OD III, 396–7). Bayle also regards Confucius as a model of morality despite his atheism (CPD: OD III, 397). Thus, what all atheists have in common is not a shared ontology, but rather, a detachment of moral virtue from religion. Most Western atheists reject the idea that divinity has an impact on human lives (RQP: OD III, 728). That could explain why Bayle values the moral attributes of divinity more than the metaphysical ones (Paganini 2009, 396). Therefore, despite the differences, Chinese and Western cultures are in many ways similar. According to Bayle, the outcome of an action, rather than the principles guiding it, is what is crucial in moral issues, which can explain how different philosophical frameworks can result in positive moral results.

However, Gianluca Mori suggests that the historical existence of atheists is not the focus in Bayle's work, but rather the creation of a philosophical myth of *elite atheism*. This myth presents atheism as a superior stage of religious belief, as it involves a rational critique of religion, but is not proposed for the general population. Mori maintains that Bayle is not an enlightened philosopher who seeks to educate the people in the rejection of religion (2011, 59–60). In fact, says Mori, Bayle argues that religion is so well adapted to the passions that it is practically impossible to eradicate and manifests itself in many ways, like an endemic disease. Then the only solution to such a disease would be tolerance, not atheism. Nevertheless, authors such as Corinne Bayerl argue that following Mori's interpretation means accepting that one can apply the idea of progress to Baylean religious belief, something that makes no sense to her (2012, 23).

In our opinion, it is necessary to differentiate between three forms of atheism: *private* or *individual* atheism, *public* or *political* atheism, and *philosophical atheism*. Discriminating between these concepts is crucial in understanding how to apply the concept of progress to atheism. It is certain that Bayle's works show a shift in focus from individual atheism to public atheism. In the *Pensées diverses*, Bayle concentrated on the debate concerning the veracity of religion and its impact on individual atheism, idolatry, and superstition.[24] However, beginning with the *Continuation des Pensées diverses* in 1704, the discussion turned to the foundations of atheism as a social bond: "We no longer compare paganism and atheism except in relation to the temporal well-being of societies".[25] While the government of atheists, or public atheism, to which Mori refers could be the subject of progress, individual atheism cannot. When Bayle suggests that there are many more atheists

---

[24] We will skip the subject of virtuous atheism, as its complexities would take us away from our current discussion.

[25] "Nous ne comparons plus depuis ce lieu-là le Paganisme, et l'Athéisme que par rapport au bien temporel des Sociétez" (CPD: OD III, 376).

in Europe than is commonly believed, but that they are not known because they do not dare to publicly reveal their beliefs for fear of reprisals,[26] he is referring to the conflict between private atheism and the public theism within a Confessional State (State church). These techniques of dissimulation were used by Jewish Marranos to escape the Spanish Inquisition and by French Huguenots to avoid persecution by Louis XIV (PD: OD III, 122). Nonetheless religious dissimulation, or *Marranism*, applies to both believers and atheists and involves a distinction between private belief and public behaviour. The essential difference is that in China there is no political obligation to accept any specific dogma that affects private beliefs, as there was in the Iberian Peninsula where the Marranos lived, or in Catholic France. In China, all that is required of everyone is participation in certain public ceremonies. Thus, Confucian philosophers, who hold positions in the government in China, mask their atheism from the population for the sake of political loyalty, not out of fear. When a specific dogma becomes mandatory for the population, that's when loyalty becomes confused with fear. This is because publicly revealing one's opposition to the official religious confession of the country in which one lives can have serious consequences for public order, which Bayle analyses in his discussion of the link between sovereignty and sedition (García-Alonso 2017).

Therefore, while dissimulation is possible in terms of public behaviour, it makes no sense to apply this concept to individual belief or disbelief. An atheist and a religious person, whether they hold heterodox or orthodox beliefs, are in some way equal in the eyes of the sovereign, who should only require them to exhibit political loyalty, not to share his religious beliefs. Hence, John Christian Laursen is correct in stating that Bayle is not an elitist in religious matters (2010). Consequently, political tolerance could take the form of a State with an established religion—State religion—tolerant of other religious confessions, a State with a tolerant civil religion, or a State with no public religion at all – a non-confessional State.[27] Those systems are opposed to a Confessional State (State church) in which following the religion of the sovereign is mandatory for the entire population. The Netherlands, England during the reign of James II, and France during the Edict of Nantes all exemplify tolerant State-religion systems. Certain periods in the history of Rome and Greece illustrate tolerant civil religious systems. It is possible that Bayle viewed China as a model of non-confessional State; although Confucian civic ceremonies were required, he regarded them as civic rituals rather than religious ceremonies.

---

[26] "out ce qui vous resteroit donc seroit de dire, qu'il n'y a point de tels Athées, et qu'il est impossible qu'il y en ait. Il est vray qu'il y en a peu qui se declarent dans les pays des Chrestiens, ou des Mahometans, parce qu'il y en a peu qui se veulent exposer à estre brulez" (CPD: OD III, 399).

[27] We choose to use the expression *non-confessional State* instead of the term *Secular state* because while the former may accommodate certain privileges to specific religious denominations –such as Catholicism currently in Spain or Confucianism in Bayle's understanding of China– the latter suggest the absence of any political privileges to any confession and some level of social secularization –such as French laïcité–.

That was the outcome of adopting the Jesuit characterization of Confucianism as a moral and civic theory (Zoli 1989, 471).[28]

The idea of progress advocated by Mori can be applied to China as a non-confessional State (public atheism). In this sense, the philosopher of Rotterdam may have thought that the State religion model was a less stable political solution than a society ruled by atheist philosophers with a non-confessional State. But it seems questionable to suggest that China was the model of a post-Christian society that Bayle aspired to, as Weinstein has argued (1999, 215). After all, Bayle was aware of the importance of his historical context and sought solutions to the moral, political, and religious problems of his own time, rather than trying to design the future. It should also be noted that, maybe apart from the *Dictionnaire*, Bayle's works were not theoretical works. And the *Dictionnaire* aims to demonstrate the errors found in other dictionaries, not to present a novel catalogue of truths. Contrary to the viewpoint of Weinstein, we hold that Bayle shows no interest in constructing a philosophical program that would result in socio-political improvement. Furthermore, philosophy necessitates that it be consistently contextualized within its historical and polemic framework. Thus, Bayle's works were written in the context of confessional polemics, and in this context, a state religion with religious pluralism was likely a more realistic solution for Europe, as it already existed in Holland, was present in France under the *Edict of Nantes,* and could have persisted in England if not for the influence of the Calvinists led by Pierre Jurieu and the Orange family in the *Glorious Revolution.* Although the superiority of either model of the relationship between religion and politics is open to debate, it's important to note that a non-confessional State does not necessarily imply the superiority of private atheism in terms of morality or intelligence (Gros 2004a). Similarly, toleration does not require an atheist State and may even coexist with the exclusion of certain ideas promoted by atheists if they pose a threat to public order (Laursen 1998). As a result, we don't think that only a moral and rational atheism can provide the universal basis for true tolerance (Gros 2004b, 432). Political tolerance can coexist with the idea that the actions are driven by passion, and therefore with a society of believers. We also know that Bayle learned from the Jesuits that the Chinese people practiced both Buddhism and Taoism. Consequently, he did not confuse an atheist government with a secular State, which contradicts Gros's interpretation.

The huge difference between Chinese and European societies is that in China, the rulers are atheist philosophers. Therefore Mori argues that atheism represents a higher level of religious belief due to its rational evaluation of religion. However, we believe that statement is only true if we compare theology to philosophy, as we have proven in another paper (García-Alonso 2021). For instance, the debate on evil did not involve a dispute between reason and religion, but rather just requires the abandonment of the use of reason in the interpretation of religious beliefs. Then, we agree with Adam Sutcliffe's observation that Bayle's rejection of Spinoza's

---

[28] Bayle followed Le Gobien, and Le Gobien's thesis can be considered a simplification of Ricci's work.

philosophical system was driven by a concern about dogmatism and its monopolization of the truth (2008). In this sense, Bayle adhered to a philosophical skepticism that enabled him to abstain from any dogmatic constructions, whether theological or philosophical in nature.[29]

4.5 Conclusion

As we have seen, China plays an essential role in Bayle's philosophical work. Although the concept *society of atheists* also refers to the inhabitants of Africa and America, China is the quintessential example of how a country can be governed without relying on religious legitimization. This point is significant because it provides an example of a successful civilization that thrived without religion being a primary political or social force. Bayle is using China to demonstrate that there is no necessary connection between religion and a successful society. In contrast, he argues that the two are separate domains that should not be intertwined. Moreover, Bayle notes that although religion may have been useful in the past, it is not necessary for the flourishing of highly developed civilizations such as China. This view is consistent with Bayle's broader philosophy of separating religion from politics. He suggests that the political sphere should be guided by the law or by the promise of honour and glory that comes with conformity to social norms –social pressure– rather than religious dogma.

The idea of political tolerance is also essential in this argument. Bayle notes that China embodies political tolerance by not imposing any obligations to adhere to a particular doctrine or participate in a specific religious ceremony. This type of toleration is a religious-political model where beliefs and rituals, the two fundamental elements of any religion, are not mandatory for the entire population. For that reason, China is not simply a backdrop for the creation of the philosophical myth of atheism. Furthermore, Bayle's argument is not only significant for its historical and philosophical implications but also has contemporary relevance. Today, there are ongoing debates about the role of religion in politics and the degree to which it should influence public policy. Bayle's argument provides a crucial perspective that can help inform these debates, highlighting the importance of separating politics and religion and promoting political tolerance.

China offers us a palpable example that many of the ideas and theories developed by European philosophers in the seventeenth century could not have taken shape if they had not had other cultures in sight. The idea that Europe is fundamentally based on Christian traditions overlooks these simple facts.

---

[29] We cannot explore here Bayle's skepticism. However, one may consult the works of Paganini (1980), Maia Neto (1999), Solère (2004), Laursen (2011) and Hickson (2016).

Bibliography

Bayerl, Corinne. 2012. Primitive atheism and the immunity to error: Pierre Bayle's remarks on indigenous cultures. In *Religion, ethics, and history in the French long seventeenth century*, ed. W. Brooks and R. Zaiser. Oxford: Peter Lang.

Bayle, Pierre. 1727–1731. *Oeuvres diverses de Mr Pierre Bayle, professeur en philosophie et en histoire à Rotterdam*, 4 vols. La Haye.

———. 1740. *Dictionnaire historique et critique*, 4 vols., 5th ed. Amsterdam/Leyde/La Haye/Utrecht: P. Brunnel et al.

———. 2000. In *Various thoughts on the occasion of a comet*, ed. Robert Bartlett. Albany, NY: State University of New York Press.

———. 2014 [1686]. Pierre Bayle's The Condition of wholly Catholic France under the reign of Louis the Great, ed. Charlotte Stanley and John Christian Laursen. *History of European Ideas* 40 (3): 312–359.

———. *The Correspondance de Pierre Bayle*: http://bayle-correspondance.univ-st-etienne.fr/?lang=fr.

Bianchi, Lorenzo. 2011. Bayle e l'ateo virtuoso. Origine e sviluppo di un dibattito. In *I filosofi et la società senza religione*, ed. M. Geuna and G. Gori, 61–80. Boulogna: Società il Mulino.

Busquets, Anna. 2015. "Más allá de la Querella de los Ritos": el testimonio sobre Cina de Fernández de Navarrete, *Anuario de Historia de la Iglesia* 25, 229–250.

Charnley, Joy. 1990. Near and Far East in the works of Pierre Bayle. *The Seventeenth Century* 5 (2): 173–183.

Daston, Lorraine. 2011. Empire of observation, 1600–1800. In *Histories of scientific observation*, ed. Lorraine Daston and Elizabeth Lunbeck, 81–113. Chicago: University of Chicago Press.

García-Alonso, Marta. 2017. Bayle's political doctrine: A proposal to articulate tolerance and sovereignty. *History of European Ideas* 43 (4): 331–344.

———. 2019. Tolerance and religious pluralism in Bayle. *History of European Ideas* 45 (6): 803–816.

———. 2021. Persian theology and the checkmate of Christian theology: Bayle and the problem of evil. In *Visions of Persia in the age of Enlightenment*, ed. C. Masroori, W. Mannies, and J.C. Laursen, 75–100. Liverpool: Oxford University Studies in the Enlightenment.

Gernet, Jaques. 1982. *Chine et Christianisme. Action et réaction*. Paris: Gallimard.

Gros, Jean Michel. 2004a. Bayle et Rousseau: société d'athées et/ou religion civile. In *Pluralismo e religione civile. Una prospettiva storica e filosofica*, ed. G. Paganini and E. Tortarolo, 124–138. Milano: Mondadori.

———. 2004b. La tolerance et le problème théologico-politique. In *Pierre Bayle dans la République des Lettres: Philosophie, religion, critique*, ed. Antony McKenna and Gianni Paganini, 129–170. Paris: Honoré Champion.

———. 2014. Le 'désenchantement' du politique chez Pierre Bayle. In *Pierre Bayle et le politique*, ed. Xavier Daverat and Antony McKenna, 175–186. Paris: Honoré Champion.

Hashimoto, Keizo, and Catherine Jami. 2001. From the *Elements* to the calendar reform: Xu Guangxi's shaping of mathematics and Astrnomy. In *Statecraft and intellectual renewal in late Ming China: The cross-cultural synthesis of Xu Guangqi (1562–1633)*, ed. Catherine Jami, Peter Engelfriet, and Gregory Blue, 263–278. Leiden: Brill.

Hickson, Michael. 2016. Disagreement and Academic skepticism in Bayle. In *Academic skepticism in early modern philosophy*, ed. S. Charles and P.J. Smith, 293–317. Switzerland: Springer.

Kow, Simon. 2017. *China in early Enlightenment political thought*. New York: Routledge.

Lach, Donald F. 1977. *Asia in the Making of Europe*. A Century of Wonder: The Literary Arts, vol II-II, Chicago, University of Chicago Press, 292–297.

Lach, Donald F. 1998. *Asia in the Making of Europe*. Trade, Missions, Literature, vol III-1, Chicago, University of Chicago Press, 223.

La Mothe le Vayer, François. 1642. *De la vertu des Payens,* 278–190. https://gallica.bnf.fr/ark:/12148/btv1b86268333/f9.item.

Laursen, John Christian. 1998. Baylean liberalism: Tolerance requires nontolerance. In *Beyond the persecuting society: Religious toleration before the Enlightenment*, ed. J.C. Laursen and C.J. Nederman, 197–215. Philadelphia: University of Pennsylvania Press.

———. 2010. Son los cosmopolitas ilustrados elitistas? Reflexiones sobre la República de las Letras de Pierre Bayle. In *Cosmopolitismo y nacionalismo. De la Ilustración al mundo contemporáneo*, ed. G. López-Sastre and V. Sanfélix, 15–32. Valencia: Universidad de Valencia.

———. 2011. Skepticism against reason in Pierre Bayle's theory of toleration. In *Pyrrhonism in ancient, modern, and contemporary philosophy*, ed. D.E. Machuca, 131–144. London/New York/Dordrech: Springer.

Laven, Mary. 2011. *Mission to China: Matteo Ricci and the Jesuit encounter with the East*, [Epub]: London Faber and Faber.

Lennon, Thomas M. 2002. What kind of a skeptic was Bayle? *Midwest Studies in Philosophy* 26 (1): 258–279.

Maia Neto, José R. 1999. Bayle's Academic skepticism. In *Everything connects: In conference with Richard H. Popkin: Essays in his Honor*, ed. James E. Force and David S. Katz, 263–276. Leiden: Brill.

Meynard, Thierry, ed. 2011. *Confucius Sinarum Philosophus (1687). The first translation of the Confucian classics*. Rome: Monumenta Historica Societatis Iesu.

Minamiki, George. 1985. *The Chinese rites controversy from its beginning to modern times*. Chicago: Loyola University Press.

Mori, Gianluca. 2004. Atéisme et philosophie chez Bayle. In *Pierre Bayle dans la République des Lettres: Philosophie, religion, critique*, ed. Antony McKenna and Gianni Paganini, 129–170. Paris: Honoré Champion.

———. 2011. Religione e politica in Pierre Bayle: la 'società di atei' tra mito e realtà. In *I filosofi et la società senza religione*, ed. M. Geuna and G. Gori, 41–60. Boulogna: Società il Mulino.

Mungello, David Emil. 1985. *Curious land: Jesuit accommodation and the origins of sinology*. Stuttgart: F. Steiner Verlag Wiesbaden.

———. 2009. *The great encounter of China and the West, 1500–1800*. New York/Toronto/UK: Rowman & Littlefield Publishers.

Paganini, Gianni. 1980. *Analisi della fede e critica della ragione nella filosofía di Pierre Bayle*. Firenze: Universitá degli studi di Milano.

———. 2009. Pierre Bayle et le statut de l'athéisme sceptique. *Kriterion* 120: 391–406.

Pinot, Virgile. 1932. *La Chine et la formation de l'esprit philosophique en France (1640–1740)*. Paris: Paul Geuthner.

Sola, Diego. 2018. *El cronista de China: Juan González de Mendoza, entre la misión, el imperio y la historia*. Barcelona: Universidad de Barcelona.

Solère, Jean-Luc. 2004. Bayle, les théologiens catholiques et la retorsion stratonicienne. In *Pierre Bayle dans la République des Lettres : Philosophie, religion, critique*, ed. Antony McKenna and Gianni Paganini, 129–170. Paris: Honoré Champion.

Sutcliffe, Adam. 2008. Spinoza, Bayle, and the Enlightenment politics of philosophical certainty. *History of European Ideas* 34 (1): 66–76.

Weinstein, Kenneth Roy. 1992. *Atheism and Enlightenment in the political philosophy of Pierre Bayle*. Dissertation. UMI: Harvard University.

———. 1999. Pierre Bayle's atheist politics. In *Early modern skepticism and the origins of toleration*, ed. A. Levine, 197–223. Lanham: Lexington Books.

Weststeijn, Thijs. 2007. Spinoza sinicus: An Asian paragraph in the history of the radical Enlightenment. *Journal of the History of Ideas* 68 (4): 537–561.

Zoli, S. 1989. Pierre Bayle e la Cina. *Studi francesi* 33: 467–472.

Zürcher, Eric. 2001. Xu Guanqi and Buddhism. In *Statecraft and intellectual renewal in late Ming China: The cross-cultural synthesis of Xu Guangqi (1562–1633)*, ed. Catherine Jami, Peter Engelfriet, and Gregory Blue, 155–169. Leiden: Brill.

Chapter 5
Bayle and Japan

Fernando Bahr

Abstract The religion and customs of Japan were of great interest to men of letters in the seventeenth and eighteenth centuries, as is shown by the long article devoted to the subject by Louis Moréri in his famous *Grand Dictionnaire historique* (1674, 1st edition). In the present study, our attention is focused on another *dictionnaire*, as famous or even more famous than the first: the *Dictionnaire historique et critique* by Pierre Bayle, whose second edition (1702) includes an article on the subject. In it, Bayle describes and analyzes a set of teachings coming mainly from Buddhism, but he does so, in our view, mainly as a pretext to discuss other surrounding doctrines that irritated or annoyed him. We will see that this occurs particularly in connection with two scandalous intellectual events that caused heated controversies towards the end of the seventeenth century: (a) Spinoza's philosophy, or rather the mode of thinking called "Spinozism" that Bayle detects throughout history, and (b) the mystical movement that arose from the writings of Miguel de Molinos receiving, from the Roman Church, the denomination of "Quietism".

Keywords Japan · Buddhism · European view · Spinozism · Quietism

---

This paper was made possible by the research project *Contra la ignorancia y la superstición: las propuestas ilustradas de Bayle y Feijoo* (PID 2019-104254GB-100) financed by the Spanish Ministry of Education and Science.

F. Bahr (✉)
Instituto de Filosofía "Ezequiel de Olaso", Consejo Nacional de Investigaciones Científicas y Técnicas (CONICET), Buenos Aires, Argentine
e-mail: fernandobahr@gmail.com

© The Author(s), under exclusive license to Springer Nature Switzerland AG 2024
M. García-Alonso, J. C. Laursen (eds.), *The Importance of Non-Christian Religions in the Philosophy of Pierre Bayle*, International Archives of the History of Ideas Archives internationales d'histoire des idées 251, https://doi.org/10.1007/978-3-031-64865-6_5

5.1 Introduction

As Marta García-Alonso shows in this volume, China is a main character in the philosophical and theological scene of the 17th and 18th centuries. Without intending to be exhaustive and concentrating on the aforementioned period, "l'empire du Milieu" (China) was the object of great attraction for philosophers such as Malebranche, Leibniz, and Voltaire, for men of letters such as Nicolas Fréret and Boyer d'Argens, for theologians such as Álvarez Semedo and Louis Le Comte. Its antiquity, its state model and, for better or for worse, its amazing religious practices, stimulated the imagination of some thinkers to the point of conceiving that reign as the dreamed (or cursed) land from whose knowledge one could better understand and value the beliefs of the place where one inhabited.[1]

The case of Japan is a bit different. This country did not feed the imagination of Europe to the same extent and, according to Shin-Ichi Ichikawa (1979), appeared to the curious eyes of the seventeenth and eighteenth centuries somewhat in the shadow of the Chinese civilization and even as an extension of it. Ichikawa adds, however, that such a secondary position did not mean disinterest, far from it. Proofs of this are a rather long article dedicated to this country by the Jesuit Louis Moréri in his *Grand Dictionnaire historique* (1702; first edition: 1674), the article "Japan" added by Pierre Bayle to the second edition of the *Dictionnaire historique et critique* (1702), and the 65 articles related to various aspects of Japanese culture that can be found in Diderot and d'Alembert's *Encyclopédie* (Nakagawa 1992), 40 of which are directly or indirectly linked to Buddhism (Hoffmann 2018).

Buddhism was indeed the most compelling issue for modern European philosophers and theologians who took an interest in Japan. Despite being a doctrine born in India and whose place of greatest diffusion was China, its name, mainly in the Zen variant, was strongly linked from the outset to Japanese culture. Urs App, who has written two important studies (2010, 2012) about the reception of Buddhist thought in the European scholarly world of the period to which we refer, opens the last of them postulating the invention by the West of an Oriental philosophy, invention that was unaware of the enormous complexity of the religious landscape in the Far East. To this appreciation, he adds two that launched the research whose result is this paper, namely: (1) that "a common 'inner' doctrine of this [supposed] philosophy (…) had an unmistakable scent of Zen Buddhism" (2012, 3); and (2) that "a core portion of the content of 'Oriental philosophy' as presented in the

---

[1] As an example of a "luminous" conception of China, we could cite this passage by Guillaume-Thomas Raynal: "The greatness of cities; the multitude of villages; the number of canals, some of which are navigable and cross the empire, and others contribute to the fertility of the land; the art of cultivating these lands; the abundance and variety of their productions; the wise and gentle exterior of the people; this continual trade in good offices, of which the countryside, the high roads give the spectacle; good order in the midst of an innumerable people, whom industry maintains in very lively agitation: all this must have surprised the Portuguese ambassador, accustomed to the barbaric and ridiculous customs of Europe" (Raynal 1780, I:193). All translations are my own, unless indicated otherwise.

seventeenth, eighteenth, and early nineteenth centuries originated in sixteenth century Japan" (2012, 18).

The Japan of the sixteenth century will occupy an important place, therefore, in the development of our work. It is towards the middle of that century (1549) that the missionaries arrived on the islands and the first encounters and the first misunderstandings took place. However, and unlike App, we are not so interested in the genealogy of what he considers—with very solid arguments—an invention, but in the crossovers that occurred from that century on with other philosophical and theological doctrines in the West, doctrines that were related to Oriental philosophy. Pierre Bayle gives us a great opportunity for that task inasmuch as, trying to be faithful in the interpretation of an idea and its logical consequences, he liked also to show inconsistencies or contradictions present in other ideas that could be linked to the first and circulated in the Republic of Letters of his day. In other words, the article "Japan" from the *Dictionnaire historique et critique*—the central object of this study—was in Bayle's pen at the same time the description and analysis of a set of teachings coming mainly from Buddhism as a pretext to discuss other surrounding doctrines that irritated or annoyed him. We will see that this occurs in particular with two scandalous intellectual events that caused heated controversies towards the end of the seventeenth century: (a) Spinoza's philosophy, or rather the mode of thinking called "Spinozism" that Bayle detects throughout history, and (b) the mystical movement that arose from the writings of Miguel de Molinos receiving, from the Roman Church, the denomination of "Quietism".[2] Our paper will be concentrated on these points. But before raising the subject, and preparing the conditions that make its treatment possible, perhaps it is convenient to start reviewing some mentions of Japan found in other writings, prior to Bayle's best-known work.

---

[2] "Quietism" is the name given by Roman Catholic theology to some contemplative practices which achieved great popularity in Spain, France, and Italy during the late 1670s and 1680s. Its origin was associated with the writings of the Spanish mystic Miguel de Molinos (1628–1696) and, in particular, with his *Guía Espiritual* (1675). Its main exponents in France were Madame Jeanne Guyon (1648–1717) and François Fénelon (1651–1715). Fénelon, later Archbishop of Cambrai, gave theological support to the mystical experiences of the young woman, a support that led to a harsh polemic with Bishop Jacques-Bénigne Bossuet. The movement, especially through Madame de Maintenon, reached an important influence at the French court of Louis XIV. It was condemned as heresy by Pope Innocent XI in the papal bull *Coelestis Pastor* of 1687. In this bull, the pope listed more than 40 heretical propositions, the first three of which were the following: "1. It is necessary that man reduce his own powers to nothingness, and this is the interior way. 2. To wish to operate actively is to offend God, who wishes to be himself the sole agent; and therefore it is necessary to abandon oneself wholly in God and thereafter to continue in existence as an inanimate body. 3. Vows about doing something are impediments to perfection". The entire document can be consulted at https://www.papalencyclicals.net/innoc11/i11coel.htm. On quietism, see Armogathe (1973).

5.2 Japan Before the *Dictionnaire*, Two Examples

Whoever goes through the pages of the *Pensées diverses sur la comète* (1683) will find three references to supposed practices or customs of the Japanese. The first is found in sect. 19, titled "On the Belief in Astrology Among the Infidels of Today". There, Bayle recalls that, according to the *Relations* of the famous traveler Jean-Baptiste Tavernier (1679, 86–87), in Japan "it matters a great deal for the durability of a building, and for the good fortune of those who are to live in it, that when one begins to build it, some kill themselves out of regard for the enterprise" (PD: OD III, 20. Bayle 2000, 36).[3] The second is in section 73, "On the Abominable Idolatry of Today's Pagans", and the source is the *Relation de l'empire du Japon, comprise dans la réponse que François Caron, président de la Compagnie hollandaise en ces pays, fit au sieur Philippe Lucas* published in Paris (1664) by the VOC ("Vereenigde Oost-Indische Compagnie"). The custom referred to is as striking as the first, namely, that the Japanese, like the Indians and the Chinese, "find themselves in the most frightful deviations that can be spoken of in the matter of religion; that they worship monkeys and cows; that they consult the demon in burning mountains…" (PD: OD III, 48. Bayle 2000, 91).[4] The third, finally, is in sect. 132, whose title is "That Idolaters Have Surpassed Atheists in the Crime of Divine Lese-Majesty". It comes from a writing prepared by the *Ambassade de la Compagnie des Indes des Provinces Unies* and affirms that the Japanese, a bit like the ancient Romans, "have 365 idols intended to keep watch over the emperor's person which are placed on guard in turn, each to be on duty for a whole day. If some evil befalls the prince, they blame the day's idol, whip it or strike it, and banish it from the palace for a hundred days" (PD: OD III, 85. Bayle 2000, 163).[5]

---

[3] "On s'imagine dans le Japon qu'il importe beaucoup, pour la durée d'un edifice et pour le bonheur de ceux qui doivent y demeurer, que lorsque on commence de le bâtir, quelques-uns se tuent eux-mêmes en consideration de cette entreprise". Bayle's texts are cited following the electronic edition of his complete works published by Garnier, following the *Oeuvres diverses de Mr. Pierre Bayle, professeur en philosophie et en histoire à Rotterdam* (La Haye: P. Husson et al., 4 vols., 1727–1731). These works are cited using the initials of the work, followed by the volume in which it is found in the *Oeuvres diverses* (OD) and the page. PD: *Pensées diverses écrites à un Docteur de Sorbonne, à l'occasion de la comète* (1683); CP: *Commentaire philosophique sur ces paroles de Jésus-Christ: contrain-les d'entrer* (1686); CPC: *Continuation des Pensées diverses écrites à un Docteur de Sorbonne* (1705). As for the *Dictionnaire Historique et Critique* (DHC), we have used the fifth edition (Amsterdam / Leyde / The Hague / Utrecht: P. Brunel et al., 4 vols., 1740) citing: DHC, title of article, remark. For the PD, I use Bartlett's translation (2000).

[4] "[Q]ui ne sait que les Indiens, les Chinois et les Japonais sont dans les plus effroyables égaremens qui se puisent dire sur le chapitre de la religion; qu'ils adorent des singes et des vaches; qu'ils consultent le démon dans des montagnes brûlantes...".

[5] "Les Japonais font aujourd'hui quelque chose de fort approchant, car ils ont trois cent soixante-cing idoles destinées à veiller sur la personne de l'empereur, les quelles on met en sentinelle tour à tour, chacune pour être en faction une journée tout entière. S'il arrive quelque mal au prince, on s'en prend à l'idole du jour, on la fouette ou on la bâtonne, et on la bannit du palais pour cent jours".

As can be seen, the three references are clearly derogatory and undoubtedly contribute to supporting one of the central theses of the *Pensées diverses*, that is, that nothing can be worse or more injurious to God than idolatry, not even atheism. Bayle does not seem to have any real interest in Japan at the moment and only uses its customs as counterexamples of what a true religion should be.

In the *Commentaire philosophique sur ces paroles de Jésus-Christ "Contrain-les d'entrer"* (1686–1687) the problem specifically addressed is another: to examine whether it is not contradictory for Catholics to complain and be outraged by the persecutions that their missionaries suffer in Japan (or in China) when they themselves persecute other religious minorities (Huguenots, Jews, Socinians, etc.) in the countries they dominate. We have found in the *Commentaire* seven mentions of Japan or the Japanese in that sense. The most interesting is the third, which is found in Chapter (or Objection) VII of the Second Part whose title is *Coercion in the literal sense cannot be denied without introducing a general tolerance. The answer to that, and that the consequence is true and not absurd*. The particular context is the tolerance due to Jews and Mohammedans. Indeed, for Bayle, if the Grand Mufti were to send missionaries to Christian countries and these missionaries were caught entering houses to convert their inhabitants, there would be no right to punish them. Such missionaries, in effect, would allege the same things that Christian missionaries allege in Japan, namely, that the zeal to make the true religion known to those who are ignorant of it and to work for the salvation of their neighbor, whose blindness they lament, has pushed them to come to make their neighbor partaker of the same enlightenment. In that case, upon hearing the answers, the only Christian and reasonable way out would be to take them to speak with priests or ministers to get them out of their error, there being no right to kill them. Now, if the latter were the decision, it would be ridiculous to consider it wrong that the Japanese did the same and condemn them for their severity. The conclusion applies as much, of course, to Mohammedans in Christian countries as it does to Christians on Japanese soil:

> and provided they do nothing against the public peace, I mean, against the obedience due to the Sovereign in temporal things, they could not even be banished with justice, neither they nor those whom they could have won by their reasons; for otherwise the pagans would have done well to drive out and imprison the Apostles, and those whom they had converted to the Gospel. We must not forget the prohibition of having double weight and double measure, nor that by the same measure with which we measure others, we will be measured (CP: OD II, 420. See Charnley 1990, 177).[6]

We are aware that the *Commentaire philosophique* admits multiple interpretations (Rex 1965; Kilkullen 1988; Mori 1999; Laursen 2001; Gros 2004; Israel 2004; Laursen 2010; Solère 2016; Bahr 2018; García-Alonso 2019; etc.). In any case, it

---

[6] "[E]t pourvû qu'ils ne fissent rien contre le repos public, je veux dire, contre l'obéïssance duë au Souverain dans les choses temporelles, ils ne mériteroient pas seulement l'exil, ni eux, ni ceux qu'ils auroient pû gagner par leurs raisons; car autrement les Païens eussent bien fait de chasser et d'emprisonner les Apôtres, et ceux qu'ils avoient convertis à l'Evangile. Il ne faut pas oublier la défense d'avoir double poids, et double mesure, ni que de la même mesure, dont nous mesurerons les autres, nous serons mesurez".

does not seem that these difficulties include Bayle's position vis-à-vis the missionaries and Catholic converts who had suffered severe persecution in Japan during the first half of the seventeenth century: if these persecutions were execrable, so were those of the Christians in the territories they controlled. Otherwise, the prohibition of using double weights and double measures to judge a matter according to whether or not we are parties involved in the conflict is seriously violated.

In Bayle's two most important works prior to the *Dictionnaire*, the ten mentions we have found would indicate that, even up to 1690, Japan was still an unknown and apparently uninteresting country for Bayle. It will take more than a decade for that perception to begin to change; indeed, as Selusi Ambrogio (2021, 80) affirms, it was not until 1702, that is, for the second edition of the *Dictionnaire*, that Bayle would have a knowledge of the Far East in general and of Japan in particular "more complex and detailed". As we have said, in the case of Japan, it will not be only because of the country itself or because of the richness of its thought, but because of the connection that he finds there with other doctrines. Let's prepare the way to reach that meeting of ideas.

5.3   Japan for the European Sixteenth Century

It must be borne in mind that Buddhism is not a native doctrine of Japan. It was successfully introduced from China into Korea around the fourth century AD, and the Japanese chronicles record that Buddhism was officially transmitted to Japan in 538 or 552 by a delegation of the King of Paekche (Korea) (Bocking 1997, 644. Cf. Renondeau 1959).[7] Confucianism followed the same route, and both religions met another, this one indigenous to Japan, it seems, which is Shintō. The confluence between the three was peaceful and fruitful, to the point that, as Bocking observes, they constituted a "syncretic blend": "For example, Shintō shrines were included in the precincts of Buddhist temples throughout Japan from earliest times. Such symbols of interdependence of Buddhas and Shintō *kami* survived until the unprecedented separation of Buddhism and Shintō by government decree in the late nineteenth century" (Bocking 1997, 657).[8]

---

[7] Also Japanese writing, based on ideograms, was imported from China, which proves, if necessary, the great influence of this country on the Land of the Rising Sun (Cf. Dubois 2012, 39).

[8] "Following the Meiji restoration of 1868, a completely new form of Shintō, retrospectively referred to as 'State Shintō', was developed by the Japanese government in a conscious effort to close the door on Japan's feudal past and unite the minds of the Japanese behind an ambitious programme of modernization and industrial expansion in order to catch up with the West. Ruthlessly separated from Buddhism, and incorporating a Confucian-style doctrine of the divine emperor as the head and the ordinary people as the body of the nation, State Shintō was vigorously propagated through schools and public institutions in a programme which nationalized the Shintō shrines and used them as vehicles for the inculcation of patriotic religious ideals and patriotic docility. Eventually, any religious or indeed secular teachings which did not conform to State Shintō were either forced to adapt or were suppressed—a quashing of dissent which finally spread through every area of Japanese life in the immediate pre-war period" (Bocking 1997, 651).

The confluence had another important consequence, which Bitō Masahide points out, limiting himself to the symbiosis between Buddhism and Shintō. For the Japanese, says Masahide, religion served two basic functions. One was concerned with satisfying the needs of daily life, the other with the individual's fate after death. "*Kami* worship, or Shinto, was largely oriented toward the former, Buddhism the latter. Once the two were identified with separate functions, a person could concurrently adhere to both" (Masahide 1991, 379).[9]

Now, classical Buddhism, which originated, as we have said, in northern India and gradually spread along the Silk Road, was a doctrine of present wisdom or enlightenment, which had little to do with the afterlife. In Japan, however, by virtue of this symbiosis with Shintō and as it spread among the population, it began to adapt to the new social conditions and, in some extent, to change its initial orientation.

More specifically, there was a significant shift in their interpretation of what salvation of the individual entailed. That is, salvation came to be understood principally as the salvation of the spirits of the dead. Therefore, greater emphasis was placed on guiding these spirits to the realm of the buddhas, and less attention was paid to the question of how the individual should seek salvation during his or her own lifetime. As a result, people came to regard the holding of funerals and masses for the dead as the main religious function of temples and priests (Masahide 1991, 380).[10]

When the first Jesuit missionaries arrived in 1549, this syncretism seemed very strange to them.[11] In any case, such exoticism was less important in missionary work than barriers related to language and communication. Ideograms consisted of ideas and practices absolutely different from those known in the Western world[12]; it was not unusual that at times the missionaries judged their purpose little less than impossible. We can quote Valignano: "for however much we learn of the language,

---

[9] "The term '*kami*' does not necessarily imply a named deity, since the primary meaning of *kami* is 'sacred'—a numinous, ambivalent and energetic quality which may attach to or inhere in a variety of objects and entities, including on occasion living or dead human beings" (Bocking 1997, 651).

[10] What became known as "Funerary Buddhism", to which Bayle dedicates some words full of irony. Masahide adds: "It is significant as well that the practice of referring to the deceased as *hotoke* seems to have become common about this time. The use of the term *hotoke*, which means buddha, indicates that people believed that the deceased would enter the realm of the Buddha as the result of the religious ministrations performed by the priest on the parishioner's behalf" (Masahide 1991, 386).

[11] The first group to arrive was led by the famous missionary Francisco Xavier, who disembarked in Kagoshima, a city located in southern Japan, where he was well received by the Lord of Satsuma (Dubois 2012: 44).

[12] The Japanese language "is so elegant and so copious", said Alessandro Valignano decades later, "that they speak in one way, write in another, and preach in a third, and there is one set of words to be employed in addressing the gentry, and another in addressing the lower classes, and there is the same sort of difference, in many cases, between the words used by children and women and the words used by men. And in their writing they have an infinite number of characters, so that none of Ours can learn to write or compose books that can be shown to anyone" (*apud* Moran 1993, 179).

and with however much effort, we still sound like children compared to them, and we never reach the stage of knowing all about their writing, and being able to write books ourselves" (*apud* Moran 1993, 178).

However, slowly and with the help of Japanese interpreters or *dōjuku*,[13] the missionaries managed to transmit the rudiments of Christianity. Taking into account the limits and difficulties, they succeeded and were honored by some local feudal lords, who converted to the new faith, thus converting their territories. Among the factors that explain this success, the religious were not the only ones and perhaps they were not even the most important. The Japanese feudal lords surely did not understand much about a God creator of heaven and earth, about paradise and hell in a life after this; but it is not impossible that they did observe with admiration the military character of the Jesuit organization. Such is the thesis of Bruno Dubois, in fact, for whom the feudal lords were interested in the support of the Jesuits in order to reduce the influence over the population of the "annoying" Buddhist sects existing at that time (Dubois 2012, 50). The explanation is plausible. Otherwise, it is difficult to understand the advance of the Jesuit campaign, an advance that Valignano lists in a letter to Roberto Bellarmino written in 1601: "107 Jesuits in Japan, besides the Jesuit bishop Luis Cerqueira; 23 houses and residences of the Society; 250 Japanese lay assistants or *dōjuku*; 300,000 Christians, of which 150,000 are in the kingdoms of Arima and Ōruma, and 80,000 reside in the kingdom of Higo" (Üçerler 2003, 351).

The presence of the Jesuits in Japan, or in some kingdoms of Japan, undoubtedly had a moment of splendor in the period between the last quarter of the sixteenth century and the first decade of the seventeenth century. It grew very quickly, but just as quickly and perhaps for the same reasons, namely the political and social influence of foreign missionaries, it collapsed. Indeed, Ieyasu Tokugawa, founder and first shogun of the Tokugawa shogunate of Japan,[14] proscribed Christianity in 1614, after which missionaries and converts who refused to apostatize were persecuted. This ban was followed a couple of decades later by the "Christian" Shimabara uprising of 1637–8.[15] Since then, "all connections with the West were prohibited, except for restricted contacts with Dutch traders, who were perceived to have no religious motives" (Bocking 1997, 654).

---

[13] I quote Valignano again: "The *dōjuku* is a young man … who helps the priest as an interpreter when he has to speak, and often in hearing confessions, … and he is indispensable when they make their visits through the villages (*apud* Moran 1993, 182).

[14] The Tokugawa shogunate ruled Japan from the Battle of Sekigahara in 1600 until the Meiji Restoration in 1868.

[15] "This event is well known in the history of Japan, because 37,000 peasants entrenched in an old castle fought valiantly against the numerically superior feudal soldiers (124,000) sent by the Government and that they were all exterminated after six months of fierce resistance. Japanese historians consider that this peasant revolt was first provoked by the weight of the royalties demanded from the local feudatories, and that it was then transformed into a revolt of exceptional magnitude, not devoid of a Christian character, because it occurred in a region where the majority of the population had been converted to Christianity a century earlier by the first European missionaries and where, despite the official prohibition of this belief, there still remained a considerable number of people who have retained some tinge of Christianity" (Ichikawa 1979, 81).

The whole episode lasted barely 65 years. They were, however, 65 years of extraordinary importance in terms of what each culture was able to learn from the other. Jacques Proust has written a marvelous study on the sense of Europe refracted by the Japanese prism (Proust 1997). Our purpose, of course, goes in reverse, concerning what Europe understood from Japan. Initially, we will do it by paying attention to the testimonies left by two missionaries.

The first name to take into account is that of Luis de Torres (c. 1510–1570), who compiled information thanks to the knowledge of Japanese that his partner Juan Fernández (1526–1576) had and sent to his superior, Francisco Xavier, then in India, in several letters. In these letters, he describes the practices of Zen monks, characterizing them as "great meditators" and holding that, in their doctrine, "there is no soul and that when man dies, everything dies, because what came from nothing returns to nothing" (App 2012, 25). These letters, says App, "are the first Western documents to mention Zen by name and to bring up themes such as 'nothingness', 'no soul', and 'no yonder' –themes that were, as we will see, bound for a brilliant career in the West" (App 2012, 25).

The Jesuits were for the first time facing monks who meditated and prayed, who shaved their heads as a symbolic way of breaking with the past, who were extremely devout, but who, nevertheless, denied the existence of God. An atheist religion? How could it be possible? What kind of religion was that? Did it deserve such a name? Even with all its errors and absurdities, paganism believed in Gods; of course, so did the monotheistic religions: Yahweh and Allah were God to their respective believers. Buddhism, on the other hand, openly denied the existence of such a Being. It was scandalous and incomprehensible to the missionaries (and to almost the entire Western world up to the present).

These and other reflections were published in 1556 under the format of a book and with the title *Sumario de los errores*. Its authors were, presumably, the same Torres and Fernández. App characterizes this publication as "extremely important in several respects". We are interested in pointing out three of them in particular: "5. It is the earliest Western source clearly distinguishing esoteric and exoteric Buddhist doctrines. 6. It contains the first detailed description of Zen practice. 7. It is the first known attempt to understand central tenets of Buddhist doctrine in terms of late medieval scholastic philosophy" (2012, 35).

Regarding the distinction between exoteric and esoteric Buddhism, a Japanese term is fundamental: *hōben* (App 2012, 36, note 15). Indeed, this term was used for teachings that were adapted to the capacity and circumstances of the recipient. The teacher knew that such ideas were false, at least with the meaning with which he transmitted them, but he also knew that the truth, or the true meaning, would be too strong a food for stomachs that were not well trained and could therefore do more harm than good. In that sense, the simple devotees did not have access to the ultimate secrets of the doctrine; they were not ready for them. They could not accept, for example, the denial of the soul and consequently of any possibility of reward or punishment in the afterlife. This point was part of what the initiate knew, but that he did not transmit to the non-initiate (Bocking 1997, 649).

Now, what was this ultimate doctrine of Buddhism, the ultimate teaching of Shaka, as the Japanese called its founder? Why did they deny the permanence of the soul and any possibility of reward or punishment after death? In this regard, the key term is *honbun*, a term that the authors of the *Sumario* understood as similar to what the Greek materialists or hylozoists had taught as the foundation of reality and that the Coimbra commentators of Aristotle had called *"caos"* or *"materia prima"*. Urs App sums it up in this way:

> the *Sumario*'s authors thus present their understanding of Shaka's cosmogony: a *honbun* (something like an eternal and non-sentient *materia prima*) forms the visible world that consists of nothing but the four elements and contains "all creatures of matter and form, called *Xiquisso*". Unlike the eternal and formless *honbun*, these creatures "grow and decrease, die and are born again" (2012, 44).[16]

Shaka's esoteric teachings thus ratified what the Buddhist monks taught their initiates: there is neither a soul nor anything that remains after death. Even more, there is no God and nothing that resembles a *creatio ex nihilo*, only an eternal chaos or *materia prima* from which all beings, like waves in water, arise by chance only to eventually dissolve again into chaos in an endless circle of birth and death. Aristotle, according to the Coimbra commentators, had affirmed that such had been the doctrine of the ancient hylozoists; therefore, hylozoism or materialism was also the deepest teaching of Zen Buddhism.

The second of the testimonies to consider comes from the pen of what was surely the great figure of the Jesuit missions in Japan. We have already named him: Alessandro Valignano (1539–1606), a, Italian nobleman and lawyer who entered the Jesuit order in 1566 under not entirely clear circumstances and who received a few years later (in 1573) the position of *Visitador* (personal delegate) of all the Jesuit missions in the East Indies: Mozambique, Malacca, India, Macao and Japan.[17] He left several writings, but the one that most concerns us here is entitled *Catechismus christianae fidei, in quo veritas nostrae religiois ostenditur, et sectae Iaponenses confutantur* (2 vols., Lisbon, 1586) and was designed for the instruction of Japanese novices and preachers.[18]

The composition process of Valignano's *Catechismus* is much more intriguing than that of the *Sumario de los errores*.[19] The author, who knew nothing of Japanese, used the information provided by native converts to Christianity as well as by Jesuit missionaries with good knowledge of Japanese and who had even learned some Chinese (Luís Fróis). However, the most extraordinary thing about this case is that,

---

[16] The passages in quotation marks are excerpts from the *Sumario*.

[17] About the biography, the personality and the missionary work of Valignano, see the very interesting Üçerler 2003, 338–366, and Moran 1993.

[18] The publication was made without the consent of the author. "Its entire structure and content leave no doubt that its purpose was to equip Japanese pupils, novices, and preaching assistants as well as European missionaries with a toolkit for the refutation of native religions along with a custom-made presentation of the Christian alternative", says App (2012, 55).

[19] This complicated web of translations and interpretations between Japanese, Portuguese, Latin and even Chinese is very well presented in App (2012, 54–64).

thanks to the fragments found in Evora, several passages of the *Catechismus* can be compared with the items in Japanese from where Valignano took them.[20] We will not stop here at this point, but we will mention some of the conclusions that Urs App derives taking it into account.

It must be understood that what Valignano was most interested in were the esoteric teachings of Buddhism. In them was the core that had to be fought. Lacking the elements to understand, for example, the philosophy of non-duality[21] and counting only on the elements that his scholastic training gave him, he interpreted in ontological terms what the Buddhist monks taught as the path of enlightenment. His conclusion was that the two doctrines were in almost perfect opposition. If for the Christian teachings God is creative, powerful and intelligent, for the Buddhist ones the principle ("One Mind" o "Buddha nature") lacks all that. If for the Christian doctrine God is provident, for the Buddhist doctrine the Mind remains aloof from the world. If the Christian God is transcendent and can never be confused with any of his creatures, the Buddhist one is identical with all things. If the Christian doctrine teaches that the human soul is sinful and can only be saved by grace, the Buddhist doctrine understands that the principle is identical with man's heart-mind. If the Christian doctrine teaches that the human soul can never be perfect and only finds its place of peace in the afterlife, the Buddhist doctrine affirms the possibility of perfection and of union with the first principle in this life through meditation.[22]

"The principle of everything is one and forms itself the nature and substance of things, such that all things have one and the same substance as the first principle" (*apud* App 2012, 71). "They argue that there is only a single first and supreme principle of all things that is inherent in separate things and both man's own heart and all other things do not differ at all from that principle. All that exists, whatever it may be, dissolves into one and the same principle that they call *isshin*" (*apud* App 2012, 72). "They say that there is no difference between ignorance and wisdom. To those who question them about this, they further explain that good and evil are not two and that there is no distance between right and wrong" (*apud* App 2012, 73). Ideas like these were interpreted by Valignano according to the elements at his disposal.

---

[20] The Evora fragments are a set of 68 items brushed with Japanese ink and handwriting that were discovered in 1902 in the Evora museum as part of the lining of the wooden frame of a Japanese screen. App calls them "the Rosetta stone of Oriental philosophy" because of their value in deciphering what were the sources, or part of the sources, that Valignano used as input to write his *Catechismus*. It is worth quoting a passage: "While the history of ideas teems with cases where an invention cannot be traced to a specific source and where guesswork must play a large role, our "Rosetta stone" –the Japanese fragments hidden in the Evora screen together with their Latin counterpart in Valignano's *Catechismus*– offers the chance to study the initial stages of an intercultural translation process. We can thus, as it were, examine the seed and sprouting of a flower of imagination, 'Oriental philosophy', in the sixteenth and seventeenth century" (App 2012, 20).

[21] The philosophy of "non-duality" postulates that underlying the multiplicity and diversity of experience there is a single, infinite and indivisible reality, whose nature is pure consciousness, from which all objects and selves derive their apparently independent existence.

[22] This last item is extremely important, in our opinion, in the discussion of Bayle and Quietism.

The first sounded clearly like an ontological thesis and it was difficult not to awaken in him the echoes of Ionian monism (Parmenides and Melissus) regarding the uniqueness, eternity, immobility. and immutability of being. Esoteric Buddhism, according to Valignano, held a monistic thesis that had to be refuted with the arguments that had proven so effective since ancient times: if there is but one reality, and everything else is only apparent, real changes in the world, or real relations among things, are impossible. The *isshin* concept ratified the absurd: what we consider changes should be seen as the movements of the sea, waves that suddenly ascend to suddenly descend a moment later. This must be understood not only of material things, but also of the human soul, that, far from being created immaterial and unalterable, it's a passing shadow that melts back into "the common principle of things", thus eliminating any possibility of reward and punishment in the afterlife. The third passage quoted derives the consequences. What sense does it make to separate wisdom from ignorance or good from evil? Such a distinction would be as ephemeral or unreal as the distinction between the waves of the sea.

If Valignano had understood the language of Zen better, he might also have understood that what the monks taught should not be rigorously interpreted in the terms of Western philosophy. The monks referred to a process of enlightenment or awakening (*satori*), where the mind, unhampered by dualistic thought, achieves a moment of total presence and clarity in which the ground and essence of everything becomes apparent. Only under these conditions do the distinctions between wisdom and ignorance or good and evil disappear along with every other duality of everyday life.

The summary that we have tried to make up to this point is of course insufficient to account for the difficulties that the Jesuit missionaries had when it came to understanding key aspects of the Buddhist teachings. In any case, we believe that from what has been said we have highlighted the importance of Valignano's writings and especially his *Catechismus*. The last question that remains to be answered is why those papers published without their author's knowledge in 1586 became so important to our topic. Urs App gives us the key:

> Republished as part of Possevino's *Bibliotheca selecta* ([1]1593, [2]1603), Valignano's *Catechismus* became a textbook for generations of missionaries as well as young Europeans studying at Jesuit colleges. Among the latter was Pierre Bayle who made extensive use of it in his bestselling *Dictionnaire critique* of 1702. Thus relatively obscure mission materials from faraway Japan burrowed their way into European public consciousness and ended up furnishing major buildings blocks for the invention of "Oriental philosophy" (2012, 57).

5.4 Japan in Bayle's *Dictionnaire Historique et Critique*

At first glance, an article in Bayle's *Dictionnare* bearing the name of a country seems a bit strange. Indeed, most of the articles in that book refer to philosophers, theologians, men of letters, or religious groups. This singularity, however, dissipates when Bayle announces that he will limit himself to examining "some points of the theology of these insulars" (DHC, Japan, in corp.). In this way, the article re-enters, so to speak, in the "editorial line" of the *Dictionnaire*: its subject will not be geography or history, but theology.

Now, how to orient oneself in Japanese theology or in Japanese theologies? The problem seems not easy. Bayle draws on two sources: Jean Crasset's *Histoire de l'église du Japon* (quoted from the summary provided by the *Journal de Savans* in its edition of July 18, 1689)[23] and, as has been said, the *Bibliotheca selecta* of the Jesuit Antonio Possevino, who dedicates book X of his first volume to transcribe what was said by Valignano regarding to the problem of the salvation of the Japanese and other oriental peoples.

According to the first source, in Japan there are three main sects:

> The first does not expect another life than this and does not recognize any other substance than that which reaches the senses... The second, which believes in the immortality of the soul and in another life, is followed by the most honest people and bears the name of the sect of the men of the very high God. The third is that of the worshipers of Xaca (DHC, Japon, in corp.).[24]

Bayle is interested in highlighting two other things from Crasset's book: (a) that all ministers destined for the service of the gods in Japan, that is, the bonzes, profess to live in celibacy; (b) that there are a total of 12 sects or 12 religions in Japan and everyone has the freedom to follow whichever one they please without implying confrontation or persecution. About the celibacy of the monks, he agrees with Crasset that it is often a strategy "to hide their debauches under the guise of an austere life" (DHC, Japon, A).[25] But, he adds, the same could be observed by a Japanese or a Chinese in the West: false celibates and other examples of double standard in priests.[26] About freedom of religion, he does not add any comment, but the mere quote evokes a passage from the *Commentaire philosophique* in which he describes

---

[23] Crasset published this book under the pseudonym of "M. l'abbé de T."

[24] "La premiere n'espere point d'autre vie que celle-ci, et ne connoit point d'autre substance que celle qui frappe les sens ... La seconde, qui croit l'immortalité de l'ame & une autre vie, est suivie par les plus honnêtes gens, et est appellée la Secte des hommes du Dieu très-haut. La troisieme est cell des adorateurs de Xaca".

[25] "[C]achent leurs debauches sous l'apparence d'une vie austere".

[26] "The missionaries who go to the Indies publish reports in which they spread out the falsehoods and frauds which they have observed in the worship of these idolatrous nations. They make fun of them; but they have to fear that they will be reminded of the *quid rides? Mutato nomine de te fabula narratur*; or the reproach that they deserve the reprisals to which are exposed those who, unaware of their own defects, discover with the utmost sagacity the vices of others" ("Les Missionaires qui vont aux Indes en publient des Relations, où ils étalent les faussetez et les fraudes qu'ils ont observées dans le culte de ces Nations idolâtres. Ils s'en moquent; mais ils ont à craindre qu'on ne les fasse souvenir du *quid rides? mutato nomine de te fabula narratur*; ou du reproche que méritent, et des représailles à quoi s'exposent ceux qui méconnoissent leurs défauts, et découvrent avec la derniere sagacité les vices d'autrui". DHC, Japon, A).

a similar situation in ancient Greece to conclude that it is civil intolerance, and not civil tolerance, that ruins states.[27]

According to Possevino (or, better, Valignano), on the other hand, the most general division that can be made of the sects of Japan is that "some make a profession of dwelling on appearance while others seek reality that does not reach the senses and which they call the truth" (DHC, Japan, in corp.) (les unes font profession de s'arrêter à l'apparence, et que les autres cherchent la réalité qui ne frappe point les sens, et qu'ils appellent la vérité). Bayle adds: "Those who stop at appearances admit another life after this one (C) for the eternal reward of the good people and for the eternal punishment of the wicked. But those who seek the inner and insensible reality reject heaven and hell, and teach many things that have a lot to do with Spinoza's opinion (D)" (Ceux qui s'arrétent à l'apparence admettent une autre vie après celle-ci (C), pour la récompense éternelle des gens de bien, et pour la punition éternelle des méchans. Mais ceux qui cherchent la réalité intérieure et insensible rejettent le Paradis et l'Enfer, et enseignent des choses qui ont beaucoup de rapport à l'opinion de Spinoza (D)).

Both sets of teachings interest Bayle greatly, albeit for different reasons. The first, those of the sectarians of appearance, are an example of the political use of religion. They follow the doctrine of Amida, Xaca, and Fotoque, says Bayle in remark C, but, in reality, their most important spiritual guide is found in the teachings of the latter, whom they also consider to be the author of all the laws of Japan.[28] For this reason, they believe that those who comply with the laws of Fotoque will be reborn after their death and will live forever in certain regions "in which the inhabitants are in a fullness of satisfaction that makes them enjoy a sovereign happiness" (DHC, Japan, C) (dont

---

[27] "To clearly show the absurdity of those who accuse tolerance of causing dissension in states, we need only appeal to experience. Paganism was divided into an infinity of sects, and rendered very different cults to their gods, and even the principal gods of one country were not those of another country; however, I do not remember having read that there was ever any war of religion among the pagans, except against people who plundered the Temple of Delphi, for example: to compel a people to leave its religion to take another, I see no mention of it among the authors. There is only Juvenal who speaks of two cities of Egypt which mortally hated each other because each held that only their gods were gods. Elsewhere great calm, and great tranquility; and why? Because one and the other mutually tolerated the rites" ("Pour montrer évidemment l'absurdité de ceux qui accusent la tolérance de causser des dissensions dans les Etats, il ne faut qu'en appeler à l'expérience. Le Paganisme étoit divisé en une infinité de Sectes, et rendoit à ses Dieux des cultes fort diférents les uns des autres, et les Dieux même principaux d'un païs n'étoient pas ceux d'un autre païs; cependant je ne me souviens point d'avoir lû qu'il y ait jamais eu de guerre de Religion parmi les Païens, si ce n'est contre des gens qui pilloient le Temple de Delphes, par exemple. Mais de guerre faite à dessein de contraindre un peuple à quitter sa Religion pour en prendre une autre, je n'en vois point de mention chez les Auteurs. Il n'y a que Juvenal qui parlede duex Villes d'Egypte qui se haïssoient mortellement, à cause que chacune soûtenoit qu'il n'y avoit que ses Dieux qui fussent des Dieux. Par tout ailleurs grand calme et gran tranquillité; et pourquoi? Parce que les uns toléroient les rites des autres". CP: OD II, 363–364).

[28] Following Possevino, Bayle seems to take as different divinities what are but different personifications of the Buddha. Amida or Amitabha refers to the Buddha revered in a very popular school of Mahayana Buddhism called Pure Land Buddhism; Shaka or Xaca is the Japanese name for the historical Buddha Shakyamuni; Hotoke or Fotoque, finally, are names linked to "Funerary Buddhism", i.e., as we have seen, the Buddha celebrated at funerals and masses for the dead (Cf. *supra*, note 11).

les habitans sont dans une plénitude de satisfaction qui les fait jouïr d'une souveraine félicité). Those who transgress them, on the other hand, "will pass from this life to certain infernal places and will suffer six kinds of pains of which they will never see the end" (DHC, Japan, C) (ils passeront de cette vie en certains lieux infernaux, et ils y souffriront six sortes de peine, dont ils ne verront jamais la fin).

What to think of this sect? We already know the distinction between exoteric and esoteric doctrines and we have seen what the term *hōben* meant as an expedient to keep calm those who were not prepared for stronger doctrines. These teachings, therefore, do not even deserve to be refuted as they are nothing more than a tale for weak spirits, as the bonzes themselves recognize. Bayle quotes Possevino: "The bonzes themselves obviously admit that this whole system of *Camus* [*Kami*] and *Fotoque* was built, or rather forged, in favor of the ignorant and imbeciles" (DHC, Japan, C).[29]

The only refutation worth undertaking, therefore, is that of the esoteric or inner doctrine. Before that, however, it is interesting to remember that the idea of a "religion for the vulgar", where the fear of hell and the hope of a paradise are forged resources destined to produce mainly political obedience, was for Bayle a well-known cliché in free thought of the first half of the seventeenth century in relation to Christianity. That now he found this topic described by a Jesuit and attributed to Buddhist monks, could not help but feed his sarcasm and, we think, he could well have repeated here the passage from Horace which was mentioned a moment ago about the similarities of double standards between bonzes and Christian priests.[30]

Now, beyond that sarcasm, the core of the discussion for Bayle was not here but in the esoteric teachings, as Valignano already knew. Even more so when Bayle could add one more monist to the numerous series that had been nurtured since ancient times: the prince of them all, Baruch Spinoza. Remark D is devoted to this discussion. It starts this way:

> They [the practitioners of esoteric teachings] neglect the exterior, they apply themselves only to meditation, they send away all discipline which consists of words, they attach themselves only to the exercise which they call *Soquxin Soqubut*, that is to say, *the heart*. They assure that there is only one principle of all things, and that this principle is found everywhere and that the heart of man and the interior of other beings do not differ from this principle and that all beings return to this common principle when destroyed (DHC, Japan, D).[31]

---

[29] "Les Bonzes eux-mêmes avouent manifestement que tout ce Système de Camus et de Fotoque a été bâti, ou plutôt forgé en faveur des ignorans, et des esprits imbecilles".

[30] Cf. *supra*, note 27.

[31] "Ils négligent l'extérieur, ils s'appliquent uniquement à méditer, ils renvoient au loin toute discipline qui consiste en paroles, ils ne s'attachent qu'à l'exercice qu'ils appellent SOQUXIN SOQUBUT, c'est-à-dire *le coeur*. Ils assurent qu'il n'y a qu'un principe de toutes choses, et que ce principe se trouve par-tout, & que le coeur de l'homme, et l'intérieur des autres êtres, ne diffèrent point de ce principe, et que tous les êtres retournent à ce principe commun quand ils sont détruits". Behind Possevino is Valignano, we know. Luckily, thanks to Urs App, we have the passage from Valignano where the aforementioned Japanese expression (*Soquxin Soqubut* or *sokushin sokubutsu*) appears. This is the complete passage of Valignano: "Third: The innermost heart of man is one and the same thing as the first principle of things. When people pass away, their hearts also completely perish and disappear; yet the first principle in them that had given them life persists. From this follow according to this doctrine that there is no life after the present one, and there is no recompense for the good or punishment for evildoers" (*apud* App 2012, 83). And this is the corresponding passage in Japanese of the Evora screen from where Valignano got his interpretation: "3) They teach

A moment later, and always following Possevino, the doctrine is summarized in four points: (1) that there is only one principle of all things; that this principle is supremely perfect, that it is wise, but that it understands nothing, and takes no heed of the affairs of this world; (2) that this principle is in all particular beings, and that it communicates its essence to them, so that they are the same thing as it, and that they return to it when they finish; (3) that the heart of man does not differ from this common principle of all beings, and that when men die, their hearts perish and are consumed; but that the first principle which gave them life before, still subsists in them, whence it follows that there is neither paradise nor hell, neither rewards nor pains after this life; (4) that man can, in this world, rise to the condition and supreme majesty of the first principle, since by way of meditation he can know it perfectly, and thus attain the sovereign tranquility which this principle enjoys in himself.[32]

The similarities with Spinoza's philosophy are found, of course, in the idea that the first principle of all things and all beings that make up the universe are nothing more than one and the same substance as God: "that all things are God, and that God is all things, in such a way that God and all things that exist are but one and the same being" (DHC, Japan, D).[33] For Bayle it is not a mere casual similarity nor does he posit any influence of Japanese religious teachings on Spinoza's system. In his opinion, it is actually a mode of thought, contradictory and horrible, but recurring: "It cannot be admired enough that an idea so extravagant and so full of absurd contradictions has been able to get into the souls of so many people so far removed from one another, and so different from each other in temperament, in education, in customs, and in talent" (DHC, Japan, D).[34]

Bayle's indignant rejection of this kind of thinking which is called, after him, "Spinozism" is well known. In the remark that we are analyzing, however, he does not renew his objections. Only observes that Valignano also pointed out those contradictions and that the bonzes reacted with indifference before them: "all their resource is to allege that it does not matter to men to inquire into the nature and force of the first principle" (DHC, Japan, D).[35] We conclude this section. Before

---

that man's one mind, as it is, is Buddha-nature, and that mind and Buddha must not be discriminated form ordinary sentient beings. When people die the mind maintains no essence and though some say that it returns to Buddha-nature, they do not speak of rebirths and hold that after death neither pleasure nor pain await us" (*apud* App 2012, 83). The differences are overwhelming and speak for themselves. Valignano's translation is a perfect example of what App calls "the Arlecchino mechanism" or "the Arlecchino style", that is, the habit of projecting what is familiar onto what is unfamiliar. The Japanese writer is not saying what the Jesuit attributes to him neither in content nor in style of thought. Valignano affirms a doctrine, the Japanese writer describes an experience, an "exercise", as Pierre Bayle says. Valignano also realized this, of course; however, ontology and morality, in the end, seem to weigh more on the whole than the awakening, fruit of meditation.

[32] Cf. Possevino 1593, I, 591–592.

[33] "[Q]ue toutes choses sont Dieu, et que Dieu est toutes choses, de telle maniere que Dieu et toutes les choses qui existent ne font qu'un seul et même être"

[34] "On ne peut assez admirer qu'une idée si extravagante, et si remplie de contradictions absurdes, ait pu se fourrer dans l'ame de tant de gens si éloignez les uns des autres, et si différens entre eux en humeur, en éducation, en coutumes, et en génie".

[35] "[T]oute leur ressource est d'alléguer qu'il n'importe point aux hommes de s'enquérir de la nature & de la force du prémier principe". "[H]ominum non interesse hujus principii vim, et naturam perscrutari inquirendo aut disputando" (Cf. Possevino 1593, I, 592).

that, however, let's pay attention to one last thing. It's in note 17, and Bayle says there: "It is a gross contradiction that Possevino should have reproached them with; because since they say that the greatest good of man comes from the perfect knowledge that he can acquire of the first principle, it is important for him to seek the origin of this first principle" (DHC, Japan, D).[36]

5.5 Buddhism, Quietism and Spinozism

Bayle's last quote puts us on the track of something that we would like to discuss and that could be called his "anti-mysticism". By mysticism could be understood the belief that union with or absorption into the Deity or the absolute, or the spiritual apprehension of knowledge inaccessible to reason, may be attained through contemplation and self-surrender. Contemplation and self-surrender, that is, an essentially passive attitude that excludes any type of intellectual apprehension: a not knowing that receives an experience inaccessible to knowledge. I think that this was what the Buddhist monks were aiming at, responding that it was not important to inquire about the nature and force of the first principle. The understanding did not matter, what mattered was the experience of enlightenment and union with the whole.

Bayle is always oblivious to this experience and on several occasions treats it contemptuously. What explanation could be found for this rejection? We conjecture that if Urs App were asked this question, his answer would emphasize Bayle's strategy to promote atheism.[37] Our interpretation is quite different and we prefer to evoke his Calvinist roots.[38]

Indeed, one of the characteristics of Calvinism is its insistence on the radical separation between Creator and creatures. Two important reasons come together at this point and explain it: (1) the radical sinfulness of human beings; (2) the central place that the notions of salvation by grace and justification by *sola fide* occupy in Calvinist soteriology. The first reason leads to excluding any mixture, confusion, or communion between God—a sovereignly perfect Being—and the radical lack that inhabits and defines the human. The second takes even further, if possible, that possibility by teaching that there is no meritorious action that can diminish the infinite distance that exists between God and his creatures, not even fasting, meditation, prayer or whatever.

---

[36] "C'est une contradiction grossière que Possevin aurait dû leur reprocher; car puis qu'ils disent que le plus grand bien de l'homme vient de la conaissance parfaite qu'il peut acquérir du premier principe, il lui importe de rechercher la nature de ce premier principe". The note, in addition to its value by itself, clearly shows us something curious: Bayle did not know that Possevino was only transcribing what was written by Valignano.

[37] Cf. App 2012, 219. His hermeneutical support is found in Jonathan Israel (2001, 2006).

[38] As far as we know, there is only one paper dealing with Bayle's controversy with quietism. Its author, Thomas Lennon (2015: 343–360), stresses Baylean Calvinism less than his skepticism. Marta García-Alonso (2021: 75–100), for her part, in a very interesting paper, argues that the idea of human imperfection in Bayle is not based on Christianity and original sin, but on its epistemology and on the power of the passions.

It is for these reasons, in our opinion, that Bayle rejected Christian mystics and Oriental philosophers alike. An article from the *Dictionnaire* is interesting in this regard, the one dedicated to Jean Taulerus. There, Bayle says that a mistake would be made with this author "if we did not distinguish him from those false mystics who taught in Christianity something similar to the errors of the Oriental philosophers (F), of which I spoke in Spinoza's article" (DHC, Taulerus, in corp.).[39] Who were these "false Christian mystics"? By way of illustration, Bayle quotes a long passage from of the famous Belgian mystic Jan Van Ruusbroec o Van Ruisbroeck (1293–1381)[40]; his true target, however, as we mentioned in the introduction to this paper, was the most scandalous mystical movement of that time, the Quietism of the Spanish Miguel de Molinos, which had very resonant figures in France (and the French aristocracy) such as Madame Guyon and François Fénelon. And, in the same way that Spinozism is a horrible and contradictory, but recurring form of thought, so too is Quietism, and both forms largely overlap. Bayle himself suggests it:

> It is surprising that these Christian mystics and these pagan philosophers were so in conformity with each other, that one would say that they had given themselves the word to spout the same follies, the ones in the East and the others in the West. What an admirable concert between people who had never seen each other, and who had never heard of each other! (DHC, Taulerus, F)[41]

Quietism therefore is also related, although it seems impossible, to Spinozism; is a form of Spinozism, we could say. In any case, it participates in the most widespread and flagrant nonsense in the history of thought. For this reason, it is not capricious that an article dedicated to Spinoza deals with Quietism in any of its variants. Bayle points it out at the very beginning:

> He was a system atheist, with a totally new method, although the background of his doctrine was common to several other ancient and modern, European and Oriental philosophers (A). Regarding the latter, just read what I refer to in remark D of the article dedicated to Japan and what I say below concerning the theology of a Chinese sect (B) (DHC, Spinoza, in corp.).[42]

---

[39] "On lui feroit tort si on ne le distinguoit pas de ces faux Mystiques qui ont enseigné dans le Christianisme quelque chose de semblable aux erreurs des Philosophes Orientaux, (F) dont j'ai parlé dans l'Article de Spinoza".

[40] The *Dictionnaire* dedicates an article to him and, interestingly enough, remark D of that article quotes the *Traité historique contenant le jugement d'un Protestant, sur la théologie mystique, sur le Quiétisme, et sur le démelêz de l'Evêque de Meaux...* by Pierre Jurieu (DHC, Ruysbroeck, D, note 14).

[41] "Il est surprenant que ces mystiques chrétiens et ces philosophes païens aient été si conformes les unes aux autres, qu'on dirait qu'ils s'étaient donné le mot pour débiter les mêmes folies, les uns dans l'Orient et les autres dans l'Occident. Quel concert admirable entre des gens qui ne s'étaient jamais vus et qui n'avaient jamais ouï parler les uns des autres!"

[42] "Il a été un Athée de Système, et d'une méthode toute nouvelle, quoique le fonds de sa Doctrine lui fût commun avec plusieurs autres Philosophes anciens et modernes, Européens et Orientaux (A). A l'égard de ces derniers on n'a qu'à lire ce que je rapporte dans la Remarque D de l'Article du Japon, et ce que je dis ci-dessous concernant la Théologie d'une Secte de Chinois (B)".

It is evidently a passage introduced in the second edition of the *Dictionnaire* (1702). Through it, Bayle seeks to expand the already long list of "Spinozists" that he had presented; these are also found in Japan, in China, and with Quietism, in the very heart of Europe! Now, the Spinozists, if they are not directly atheists, flirt with atheism. Is it possible to find in Quietism a variant of atheism? Oriental quietism teaches that such a relationship is not far-fetched: any doctrine that annuls the absolute sovereignty of God shares with atheism one of its main postulates and runs the risk of falling, knowing it or not, into such an abyss. That postulate that atheism and Quietism share was considered from the heart of all Christian churches the most terrible and unequivocal error: the possibility of a sinless state for the human soul, or, in other words, the postulate of human perfection.

We saw earlier[43] that this possibility of human perfection was part of the accusation that the Roman Church directed at the Christian quietists[44]; it was mainly because of it that they were condemned. We'll come back to this. Before doing so, however, it is necessary to return to remark B of the article "Spinoza". Bayle was referring to a Chinese sect and adds: "The name of this sect was *Foe Kiao*. It was established by royal authority among the Chinese in the year 65 of the Christian era. Its first founder was the son of King *In fan vam*, and he was first called *Xe*, or *Xe Kia*, and later, when he was 30 years old, *Foe*, that is, not man" (DHC, Spinoza, B).[45] A side note adds: "The Japanese called it Xaca" (DHC, Spinoza, B, note 24).[46] Xaca, Shaka, Shakyamuni are, in effect, the Japanese names for Buddha. Buddhists are quietists, therefore, and Bayle describes that practice.

> All those who seek true beatitude, they said, must allow themselves to be absorbed in deep meditations until they abandon any use of the intellect; thus, through consummate insensitivity, they will sink into the rest and inaction of the first principle, which is the true means to resemble it perfectly and to participate in happiness (DHC, Spinoza, B).[47]

---

[43] See *supra*, note 3.

[44] In note 40, we mention a book by Pierre Jurieu. There he affirms that "all the inner paths [the "interior way" of Quietism] tend to make a man perfect, to transform him with a perfect transformation, to unite him to God with such a singular union that he himself remains in some way God and passes, so to speak, into the divine substance" (1699: 16). Later he adds: "this mystical theology is a source of pride. Also these authors speak only of the perfects and of perfection, they have an intuitive view of God, they are all similar to glorified souls. These souls are transformed until Deification. They have no more sin and cannot have, they are entirely annihilated in themselves; they no longer have any self than that of God, they are an abyss in his essence" (1699: 76).

[45] "Le nom de cette secte est *Foe Kiao*. Elle fut établie par l'autorité royale parmi les Chinois, l'an 65 de l'ère chrétienne. Son premier fondateur était fils du roi *In fan vam*, et fut appelé d'abord *Xé*, ou *Xé Kia* (24), et puis quand il eut trente ans, *Foe*, c'est-a-dire, *non homme*".

[46] "Les Japonais le nomment *Xaca*".

[47] "[I]ls disent que tous ceux qui cherchent la véritable béatitude doivent se laisser tellement absorber aux profondes méditations, qu'ils ne fassent aucun usage de leur intellect, mais que par une insensibilité consommée, ils s'enfoncent dans le repos et dans l'inaction du premier principe, ce qui est le vrai moyen de lui ressembler parfaitement, et de participer au bonheur".

A marginal note, indispensable as always, refers to the article "Brachmanes".[48] If we follow that indication, it will be confirmed that what Bayle had in mind was the quietism of Miguel de Molinos and Madame Guyon. He says it explicitly: the doctrine preached by these Indian philosophers [the Brahmins] about "perfect indifference" or "the perfect state of the soul, in which the soul itself becomes God" is like "the favorite dogma" of quietism, for whom "true beatitude consists in nothingness". The remark continues like this:

> Both in Europe and in China these crazy visions have not ceased to be eloquently refuted, but, to the shame of our century and our atmosphere, they find fearful apologists here. Note that the dogma of the Brahmins is in a certain sense less horrifying than that of our mystics, because the first establish indifference and perfect stillness in a transformation of the soul into God, which the latter explain by ideas taken from the consummation of marriage [follows a long quote from Madame Guyon's *Commentaire au Cantique des cantiques de Salomon*, 1688].[49] The absurdity of this dogma in relation to metaphysics is monstrous, because if there is something true in our clearest ideas, it is that it is impossible with all impossibility that there is a real transformation either of God into the creature or of the creature into God (DHC, Brachmanes, K).[50]

We consider the last sentence very important to understand Bayle's position on many philosophical and theological issues. According to it, the supposed promoter of atheism would rather be a defender of the (very Christian) doctrines of perfect divine sovereignty and of original sin.[51] Are we assuming then that Bayle was an orthodox Christian? The answer is "clearly not". We assume that, remembering Hubert Bost, he was a "protestant compliqué" (Bost 2008, 83–101), but that both on

---

[48] Also, this article was included only in the second edition of the *Dictionnaire*.

[49] Metaphors of carnal love were frequent among quietists, as Lennon (2015: 344, 354) shows. Bayle was horrified by them.

[50] "On ne manque point dans l'Europe, non plus qu'à la Chine, de réfuter éloquemment ces folles visions; mais, à la honte de notre siècle et de nos climats, elles y trouvent des apologistes qui se font craindre. Notez que le dogme des brachmanes est moins affreux à certaines égards que celui de nos mystiques; car ceux-ci établissent l'indifférence et la quiétude parfaite dans une transformation de l'âme en Dieu, laquelle ils expliquent par las idées de la consommation du mariage (...) L'absurdité de ce dogme par rapport à la métaphysique est monstrueuse; car, s'il y a quelque chose de certain dans les idées les plus claire, il est impossible de toute impossibilité qu'il se fasse un changement réél, ni de Dieu en la créature, ni de la créature en Dieu".

[51] "Christians, and above all Protestants, are more obliged than others to draw this last conclusion, they who know that original sin has corrupted human nature, and that it infects it in such a way that there is no nothing remains. Darkness obfuscates the understanding, malice depraves the will. (...) St. Paul teaches us that all men are the children of the wrath of God. The Reformed Churches publicly confess at the entrance to their exercises *that we are poor sinners, conceived and born in iniquity and corruption, useless to all good, and that by our vice we endlessly and ceaselessly transgress the holy commandments of God*" ("Les Chretiens, et sur tous les Protestans, sont plus obligez que les autres à tirer cette derniere conclusion, eux qui savent que le péché originel a corrompu la nature humaine, et qu'il l'infecte de telle sorte qu'il n'y reste rien d'entier. Les ténèbres obscurcissent l'entendement, la malice déprave la volonté. (...) St. Paul nous enseigne que tous les hommes sont les enfans de la colere de Dieu. Les Eglises Réformées confessent publiquement à l'entrée de leurs exercises, *que nous sommes de pauvres pécheurs, conceus et nez en iniquité et en corruption, enclins à mal faire, inutiles à tout bien; et que par notre vice nous transgressons sans fin et sans cesse les saints commandemens de Dieu*". CPC: OD III, 220).

a personal and confessional level his Calvinist roots remained and deeply defined his philosophical and theological thought. For these reasons, and despite the sharp criticisms that Gianluca Mori (1999, 321–343) launched against Bayle's supposed Augustinianism, we think that Elisabeth Labrousse was not wrong in considering him an Augustinian without grace.[52]

We have seen that for Bayle quietism (Eastern and Western) could be included in the long list of Spinozists. Finally, we will try to see if the opposite could also be true, that is, whether Spinoza's philosophy leads, or is related, to a form of quietism. In other words, we will try to answer two questions. The first is whether Spinoza defended the possibility of a sinless state and therefore of human perfection. The second is how Spinoza conceived the union with God and whether such a union can be considered a mystical experience. In the event that the answers for both were affirmative, we would find another reason for the Spinoza-Bayle opposition.

The first requirements seem to be fulfilled. In effect, Spinoza denies the possibility of original sin and affirms unequivocally the possibility of human perfection. As for the denial, we know from the *Theological-Political Treatise* (Spinoza 2007, 65), the *Political Treatise* (Spinoza 2002, 684) and the *Letters* (Spinoza 2002, 804–836) that original sin must be rejected as an impossible fact, and the texts that speak of it must be interpreted as allegories (*historiam sive parabolam*)[53] used by the prophets and theologians to show that every action is followed by some profit or damage (Domínguez 1980, 349; Preus 2001; Polka 2007). As for the affirmation, it is part of the core of Spinoza's teachings and we could find many passages that support it. The beatitude to which the *Ethics* aims is, precisely, the possession or knowledge of the necessity of our perfection. We know this perfection as we intellectually love God, that is, when we experience the joy of knowing our individual and finite reality in its unity with the total and infinite reality.[54]

---

[52] "In the seventeenth century, the dichotomy was hardly more accentuated between Protestants and Catholics than between Augustinians—Orthodox Calvinists, Jansenists, Dominicans—on the one hand, and, on the other, these heirs of Erasmus, who were the Arminians, the partisans of the theology of Saumur, the Socinians, the Molinist Jesuits, whom their adversaries qualified with indignation as Pelagians or semi-Pelagians. Division runs through every confession" (Labrousse 2003, 19).

[53] "Hence, this one prohibition laid by God on Adam entails the whole divine law and agrees fully with the dictate of the natural light of reason. It would not be difficult to explain the whole history, or parable,of the first man on this basis, but I prefer to let it go. I cannot be absolutely sure whether my explanation agrees with the intention of the writer, and many people do not concede that this history is a parable, but insist it is a straightforward narrative" (Spinoza 2007, 65).

[54] As we said, many passages can be cited. Perhaps the most resounding is in the long Scholium to Proposition XLIX of the Second Part of the *Ethics*: "My final task is to show what practical advantages accrue from knowledge of this doctrine, and this we shall readily gather from the following points: 1. It teaches that we act only by God's will, and that we share in the divine nature, and all the more as our actions become more perfect and as we understand God more and more. Therefore, this doctrine, apart from giving us complete tranquility of mind has the further advantage of teaching us wherein lies our greatest happiness or blessedness, namely, in the knowledge of God alone, as a result of which we are induced only to such actions as are urged on us by love and piety" (Spinoza 2002, 276).

Now let's go to the second question. Was Spinoza pointing to a mystical experience? This is a very difficult and highly debated issue that cyclically reappears in the studies on Spinoza. It was treated, among others, by Melamed (1933), Hubbeling (1974), Naess (1978), Wetlesen (1979), Nadler (2003) and Van Reijen (2018).

In a very insightful and interesting article, Van Reijen analyzes the previous interpretations in this regard and wonders if Spinoza should be considered as a rational Western philosopher or as a religious or oriental mystic. She adds:

> The rationalist interpretation puts him in the line of Descartes, with his scientific method and the logical way of reasoning and writing. The religious or mystical interpretation places him in the line of Confucius, Lao Tse, Buddha, Moses and the prophets, and Paul. It emphasizes Spinoza's appreciation for intuition and interprets the intellectual love of god as a mystical experience, a mystical union with god (Van Reijen 2018, 304).

Her interpretation favors this last perspective: Spinoza is a mystic or, in any case, an "anomaly" in Western philosophy (2018, 303) and the understanding that can be had of his thought—especially with regard to the highest form of knowledge, the *amor Dei intellectualis*—gains when we incorporate the relationship with some doctrines of the Far East.

Both Spinoza and many Eastern ways of thinking recognize different kinds of knowledge, the highest form being not the rational, but the intuitive one. Spinoza calls this the perspective of eternity. Buddhism sees this as a phase of enlightenment, to see oneself as dependent on the whole and in the whole" (Van Reijen 2018, 303).

Having dealt briefly with the two questions relating to Spinoza and having answered both in the affirmative, let's go back to Bayle now. Throughout the second part of this paper we have seen that Bayle relates Spinozism to Buddhism (and Quietism) in various ways. However, and as far as we know, he never accuses Spinoza of being a "quietist". Was it because he was not very interested in quietism, as Lennon argued, or was it because charging him as an atheist no longer required a worse accusation? Both explanations are possible, but a third could also be adduced: Bayle never accused Spinoza of being a quietist because he never understood the third kind of knowledge and therefore could hardly have seen there a mystical experience. His interpretation of Spinoza as a "system atheist" blinded him, we might say, to other possibilities and it took 80 years for Western philosophy, thanks mainly to Goethe and Herder, to have a radically different perspective from Spinoza's thought.[55]

5.6 Conclusion

In post-*Dictionnaire* writings, Bayle's discussion of mystical practices in Japan (and in the West) disappears. In fact, we have only found one reference to the article "Japan"; it is in the *Continuation des pensées diverses* (CPC: OD III, 413) and in the

---

[55] On this subject, the well-documented work of M. J. Solé (2011) can be consulted.

context of a "little digression" on an English author who has acknowledged the existence of speculative atheists.[56] Clearly, other issues and other adversaries were more interesting for him. Japan, Buddhism, Quietism: perhaps they were just occasions that the *Dictionnaire* offered him to put into motion his extraordinary capacity for dialectical combat and, as we said in the Introduction, to detect inconsistencies or contradictions in some ideas that circulated in the Republic of Letters at the end of the seventeenth century and the beginning of the eighteenth century.

The Far East was and is a cultural ambit that has always aroused curiosity in the Western world: it is part of that "other" that we will never be able to fully understand. For this reason, in the first half of our paper we have brought out the conditions in which the religious conquest of Japan by Christianity took place—and in which it succumbed. Upon discovering them, thanks mainly to the works of Urs App and Selusi Ambrogio, we found it interesting to show that not only agreements but also misunderstandings produced ideas and changed practices in the genesis of the Modern world.

But our aim was not to study *per se* the frustrated religious conquest of Japan by the Jesuits, but to study it in relation to the echoes that this undertaking had aroused in one of the most widespread books in European libraries of the eighteenth century, Pierre Bayle's *Dictionnaire historique et critique*. First of all is the article "Japan" with the long excerpts from Valignano-Possevino and then the articles that the author connected, directly or indirectly, with that one. As is well known, to follow these connections is the best way to approach the *Dictionnaire* because it allows us to see, beyond the alphabetical order, how Bayle appealed to "the sagacity of the reader" to complete the idea or criticism that his own pen did not want to expose.

In the case of "Japan", that network of internal references revealed to us that Bayle was not interested so much in writing about Buddhism as in showing that some notions of those distant bonzes resembled concepts of Spinoza's philosophy or the teachings of the "quietist" mystics, that is, two highly topical intellectual events in the Republic of Letters of his time. In no case does Bayle propose direct or indirect influences: even without knowing or ever hearing from each other Zen Buddhism, Spinozism, and Quietism share, as if it were, an archetype of thought, an archetype that does not separate God from the world and assumes that somehow God and the beings of the world (specially the human beings) are or can be the same thing.

For a Calvinist like Bayle, for an "Augustinian without grace" as Labrousse called him, this thought was a gross metaphysical error and unacceptable theological folly: "if there is something true in our clearest ideas, it is that it is impossible with all impossibility that there is a real transformation of God into the creature or of the creature into God" (DHC, Brachmanes, K). Original sin was a flagrant reality before his eyes and constituted an absolutely insurmountable barrier to any project of "meeting" with God, the absolutely perfect Being. The Buddhists did not know

---

[56] The reference, in any case, is minor and only mentions the bad reputation that the Jesuits calculatingly spread about the bonzes.

original sin, and did not even grant reality to evil; the quietists thought that union with God was possible for the "perfect" who allowed themselves to be inhabited by divinity; Spinoza considered, whether or not Bayle understood him, that intuition transcended reason and became *amor Dei intellectualis*. Bayle was at the antipodes of all of them.

Bayle was a very important figure for the birth of Modernity. His work, and mainly the *Dictionnaire*, tried to gather the knowledge of his time, not with a definitive purpose—that is clear—but to make it available to the "curious public" who could not access the highest spheres of scholarly knowledge. But he took care not only to gather it, but also to correct it—in terms of facts—and criticize it—in terms of ideas—trying to free the "curious" from errors and frauds that could misguided their thinking. It is for this reason, and not because he was personally a more or less hidden atheist, in our opinion, that Bayle deserves a place in the history of the emancipation of Western man.

Bibliography

The Works of Pierre Bayle

Bayle, Pierre. *Commentaire philosophique sur ces paroles de Jésus-Christ: contrain-les d'entrer* (CP). In *Oeuvres diverses de Mr Pierre Bayle, professeur en philosophie et en histoire à Rotterdam* (OD), Vol. II, 1727–1731. La Haye: P. Husson *et al.*

———. *Pensées diverses écrites à un docteur de Sorbonne, à l'occasion de la comète* (PD). In *Oeuvres diverses de Mr Pierre Bayle, professeur en philosophie et en histoire à Rotterdam* (OD), Vol. III, 1727–1731. La Haye: P. Husson *et al.*

———. *Continuation des Pensées diverses écrites à un Docteur de Sorbonnea comète* (CPC). In *Oeuvres diverses de Mr Pierre Bayle, professeur en philosophie et en histoire à Rotterdam* (OD), Vol. III, 1727–1731. La Haye: P. Husson *et al.*

———. 1740. *Dictionnaire historique et critique* (DHC), 4 vols., 5th ed. Amsterdam/Leyde/La Haye/Utrecht: P. Brunel *et al.*

———. 2000. In *Various thoughts on the occasion of a comet*, ed. R.C. Bartlett. Albany: State University of New York.

Other Bibliography

Ambrogio, Selusi. 2021. *Chinese and Indian ways of thinking in early modern European philosophy. The reception and the exclusion.* London: Bloomsbury Academic.

App, Urs. 2010. *The birth of orientalism*. Philadelphia: University of Pennsylvania Press.

———. 2012. *The cult of emptiness. The Western discovery of Buddhist thought and the invention of oriental philosophy.* Rorschach/Kyoto: University Media.

Armogathe, Jean-Robert. 1973. *Le quiétisme*. Paris: Presses Universitaires de France.

Bahr, Fernando. 2018. El *Commentaire Philosophique* de Pierre Bayle: una interpretación alternativa. *Hispania Sacra* 70 (141): 65–73.

Bocking, Brian. 1997. The origins of Japanese philosophy. In *Companion Encyclopedia of Asian Philosophy*, ed. B. Carr and I. Mahalingam, 641–659. London/New York: Routledge.

Bost, Hubert. 2008. Pierre Bayle, un "protestant compliqué". In *Pierre Bayle (1647-1706), le philosophe de Rotterdam. Philosophy, Religion and Reception*, ed. W. van Bunge y H. Bots, 83–101. Leiden/Boston: Brill.

Charnley, Joy. 1990. Near and Far East in the works of Pierre Bayle. *The Seventeenth Century* 5 (2): 173–183.

Domínguez, Atilano. 1980. La morale de Spinoza et le salut par la foi. *Revue Philosophique de Louvain. Quatrième série* 78 (39): 345–364. https://doi.org/10.3406/phlou.1980.6094.

Dubois, Bruno. 2012. *Réalité et imaginaire, le Japon vu par le XVIIIe siècle français*. Thèse pour obtenir le grade de Docteur à l'Université de Bourgogne. Discipline: Lettres. https://theses.hal.science/tel-00843582/document.

García-Alonso, Marta. 2019. Tolerance and religious pluralism in Bayle. *History of European Ideas* 45 (6): 803–816.

———. 2021. Persian theology and the checkmate of Christian theology: Bayle and the problem of evil. In *Visions of Persia in the Age of Enlightenment*, ed. C. Masroori, W. Mannies, and J.C. Laursen, 75–100. New York: Oxford University Press; Voltaire Foundation in association with Liverpool University Press.

Gros, Jean-Michel. 2004. La tolérance et le problème théologico-politique. In *Pierre Bayle dans la République des Lettres. Philosophie, religion, critique*, ed. A. McKenna and G. Paganini, 411–442. Paris: Honoré Champion.

Hubbeling, Hubertus. 1974. Logic and experience in Spinoza's mysticism. In *Spinoza on knowing, being and freedom*, ed. J.G. Van der Bend, 126–143. Assen: Van Gorcum.

Hoffmann, Benjamin. 2018. Diderot et l'introduction du bouddhisme en occident. *Recherches sur Diderot et sur l'Encyclopédie* 53. https://doi.org/10.4000/rde.5677.

Ichikawa, Shin-Ichi. 1979. Les mirages chinois et japonais chez Voltaire. *Raison présente* 52: 69–84.

Israel, Jonathan. 2001. *Radical enlightenment. Philosophy and the making of modernity 1650–1750*. Oxford: Oxford University Press.

———. 2004. Pierre Bayle's political thought. In *Pierre Bayle dans la République des Lettres. Philosophie, religion, critique*, ed. A. McKenna and G. Paganini, 349–380. Paris: Honoré Champion.

———. 2006. *Enlightenment contested. Philosophy, modernity, and the emancipation of man 1670–1752*. Oxford: Oxford University Press.

Jurieu, Pierre. 1699. *Traité historique: contenant le jugement d'un protestant, sur la théologie mystique, sur le quietisme, & sur les démêlez de l'evêque de Meaux avec l'archevêque de Cambray, jusqu'à la bulle d'Innocent XII. & l'Assemblée provinciale de Paris du 13 de may 1699 inclusivement*. [S.l] : [s.n.].

Kilkullen, John. 1988. *Sincerity and truth. Essays on Arnauld, Bayle, and toleration*. Oxford: Clarendon Press.

Labrousse, Elisabeth. 2003. Bayle, ou l'augustinisme sans la grâce. In *La raison corrosive. Études sur la pensée critique de Pierre Bayle*, ed. I. Delpla and Ph. de Robert, 19–23. Paris: Honoré Champion.

Laursen, John. 2001. The necessity of conscience and the conscientious persecutor: The paradox of liberty and necessity in Bayle's theory of toleration. In *Dal necessario al possibile. Determinismo e libertà nel pensiero anglo-olandese del XVII secolo*, ed. L. Simonutti, 211–228. Milano: FrancoAngeli.

———. 2010. Pierre Bayle: el pirronismo contra la razón en el *Commentaire Philosophique*. *Revista Latinoamericana de Filosofía* 36 (1): 35–58.

Lennon, Thomas. 2015. Leibniz, Bayle and the quietist controversy. In *Leibniz et Bayle: Confrontation et dialogue*, ed. C. Leduc, P. Rateau, and J.-L. Solère, 343–362. Stuttgart: Franz Steiner Verlag.

Masahide, Bitō. 1991. Thought and religion, 1550-1700 (translated by K. Wildman Nakai). In *The Cambridge history of Japan*, ed. J. Hall and K. McClain, vol. 4, 373–424. Cambridge: Cambridge University Press.

Melamed, Samuel. 1933. *Spinoza and Buddha. Visions of a dead god*. Chicago: University of Chicago Press.

Moran, Joseph. 1993. *The Japanese and the Jesuits. Alessandro Valignano in sixteenth-century Japan*. London: Routledge.

Moréri, Louis. 1702. *Le grand dictionaire historique ou le melange curieux de l'histoire sacree et profane*, 4 vols. Amsterdam/La Haye: aux Dépens de la Compagnie.

Mori, Gianluca. 1999. *Bayle philosophe*. Paris: Honoré Champion.

Nadler, Steven. 2003. Spinoza and Philo; the alleged mysticism in the *Ethics*. In *Hellenistic and Early Modern Philosophy*, ed. J. Miller and B. Inwood, 232–250. Cambridge: Cambridge University Press.

Naess, Arne. 1978. Trough Spinoza to Māhayāna Buddhism or trough Māhayāna Buddhism to Spinoza. In *Spinoza's philosophy of man*, ed. J. Wetlesen, 136–158. Oslo: Universitets Forlaget.

Nakagawa, Hisayasu. 1992. *Des Lumières et du comparatisme. Un regard japonais. sur le 18e siècle*. Paris: Presses Universitaires de France.

Polka, Brayton. 2007. *Between philosophy and religion: Spinoza, the bible, and modernity*, 2 vols. Lanham/Plymouth: Lexington Books.

Possevino, Antonio. 1593. *Bibliotheca selecta qua agitur de ratione studiorum in historia, in disciplinis, in salute omnium procuranda*, 2 vols. Roma: Ex Typographia Apostolica Vaticana.

Preus, Samuel. 2001. *Spinoza and the irrelevance of biblical authority*. Cambridge: Cambridge University Press.

Proust, Jacques. 1997. *L'Europe au prisme du Japon, XVIe-XVIIIe siècle. Entre humanisme, Contre-Réforme et Lumières*. Paris: Albin Michel.

Raynal, Guillaume Thomas. 1780. *Histoire philosophique et politique des établissements et du commerce des Européens dans les deux Indes*. Genève: Jean-Léonard Pellet.

Renondeau, Gaston. 1959. La date de l'introduction du bouddhisme au Japon. *T'oung Pao, Second Series* 47 (1/2): 16–29.

Rex, Walter. 1965. *Essays on Pierre Bayle and religious controversy*. The Hague: Martinus Nijhoff.

Solé, Jimena. 2011. *Spinoza en Alemania (1670–1789). Historia de la santificación de un filósofo maldito*, Córdoba (Argentina): Brujas.

Solère, Jean-Luc. 2016. The coherence of Bayle's theory on toleration. *Journal of the History of Philosophy* 54 (1): 21–46.

Spinoza, Benedict de. 2002. *Complete works*, ed. M. Morgan, trans. S. Shirley. Indianapolis/Cambridge: Hackett Publishing Company.

———. 2007. *Theological-political treatise*, ed. J. Israel, trans. S. Silverthorne and J. Israel. Cambridge: Cambridge University Press.

Tavernier, Jean-Baptiste. 1679. *Recueil de plusieurs relations et traités singuliers et curieux qui n'ont point été mis dans ses six premiers voyages*. Paris: Gervais Clouzier.

Üçerler, Antoni. 2003. Alessandro Valignano: Man, missionary, and writer. *Renaissance Studies* 17: 3. Special Issue: "Asian Travel in the Renaissance": 337–366.

Van Reijen, Miriam. 2018. Might Spinoza be considered more as an exponent of the Oriental enlightenment, than as an exponent of the Western enlightenment? *Araucaria. Revista Iberoamericana de Filosofía, Política y Humanidades* 20 (39): 297–310.

Wetlesen, Jon. 1979. *The sage and the way. Spinoza's ethics of freedom*. Assen: Van Gorcum.

Chapter 6
Bayle's Reception of Greco-Roman Religion and Culture

Parker Cotton

Abstract This chapter examines two main areas of Bayle's reception of Greco-Roman religion and culture. (1) classical scholars as models of virtue, despite their pagan religion. (2) the use of classical religion to engage controversial points of Christian dogma. These two areas of reception connect in Bayle's writing in how moral exemplars can be found amongst the Ancients that force comparison with contemporary ecclesial claims of moral superiority. The chapter consists of a survey of articles from the *Dictionary* which engage Greco-Roman religion. The articles for this study coalesce around several common themes: the problem of providence, scandalous behaviour, false gods, and pagan virtue. The use of Greco-Roman religion as a lens to critique contemporary religious practice and belief is demonstrated. The question of virtue is central to those articles dealing with persons and religion of Greco-Roman culture. The relation to the contemporary church made explicit at times with Bayle drawing comparisons both positive ('our God is not like *that*') and negative ('Christians today behave similarly'). The comparative moves force the reader to keep these comparisons in play even when they are not made explicit. Virtue is inseparable from larger Baylean concerns around the nature of divinity and human behaviour, towards each other and towards the divine. The Greco-Roman articles allow Bayle to question contemporary religious ethics while maintaining a degree of separation. We see Bayle use Greco-Roman mythology as a commentary upon the dominant religion of his day.

Keywords Reception history · Virtuous pagan · Polytheism · Classical religion · Republic of Letters · Providence

P. Cotton (✉)
Wycliffe College, University of Toronto, Toronto, ON, Canada

Academic Registrar, Institute for Christian Studies, Toronto, ON, Canada
e-mail: parker.cotton@mail.utoronto.ca

Writing in the *Projet*, his test project for the famous *Historical and Critical Dictionary*, Pierre Bayle describes his role as the corrector of existing Dictionaries as an undertaking of Herculean proportions. For the editor of factual errors in encyclopaedias,

> His most malicious enemies could not have prescribed to him a harder task, had they over his destiny the same power that Eurystheus had over that of Hercules. It is worse than going to fight monsters it is attempting to lop off the Hydra's heads; it is at least attempting to cleanse Augeas's stables; in fine, it is, the penance which ought to be enjoined those turbulent men who have abused their leisure and the credulity of the people, to vent under the name and authority of the Revelation all sorts of Chimera's (*DHC*, A Dissertation... A Project of a Critical Dictionary).

One constant topic of discussion, and constantly in need of correcting, is the world of Greco-Roman religion.

This chapter will focus on two areas of Bayle's reception of Greco-Roman religion and culture. The first, the appeal to classical scholars as models of virtue, despite their pagan religion.[1] This includes appeals to Rome specifically as a model for the Republic of Letters, when it fits Bayle's agenda. Secondly, the use of religion, and the Olympic gods particularly, to engage controversial points of Christian dogma. Bayle cleverly ensnares his reader with scandalous material and fanciful myths before explicitly connecting a moral failing to contemporary Christian practice, or more carefully, implicitly prompting reflection on the nature of the divine. These two areas of reception connect in some of Bayle's most famous writing, detailing pagan virtue. How, despite the failings of their deities, paragons of virtue can be found amongst the Ancients that force comparison with ecclesial claims of moral superiority. In an effort to catalogue the variety of ways Bayle engages these topics, the bulk of this chapter will take the form of surveying various articles from the *Dictionary* which engage Greco-Roman religion. The articles have been collected along several common themes: the problem of providence, scandalous behaviour, false gods, and pagan virtue. My aim is to demonstrate that even in Bayle's scattershot approach to content, we can find clear and consistent approaches. The use of Greco-Roman religion as a lens to critique contemporary religious practice and belief, with some distance, emerges as one such consistent approach.

---

[1] Marenbon 2017 is an excellent survey of the 'problem of paganism', that is, reconciling the apparent virtue of pagans with knowledge of (the true) God and salvation. Bayle is briefly mentioned with Marenbon electing to focus on Catholics due to Protestants generally following Augustine in solving the problem of paganism. The Augustinian approach is, summarily, to deny that genuine virtue is displayed by pagans for true virtue can only emerge with knowledge of the true God. However, as we will unpack, Bayle problematizes the Augustinian mode and happily brings the problem of paganism to the fore throughout his writings forcing his readers to confront the cognitive dissonance that arises should they deny the virtue pagans displayed. It might be worth interrogating whether 'pagan' should be used in meta-discussion of Greco-Roman religion and other polytheistic groups. After late antiquity, 'pagan' was generally considered a pejorative used by the dominant religion. However, neo-paganism has reclaimed the term and maintaining usage in this survey avoids anachronism.

The language of Greco-Roman religion is everywhere in Bayle's writing. Most people today, in my experience, have little working knowledge of Greco-Roman myths. But these stories provided a framework for the Enlightenment educated to communicate within a shared cultural fabric. Until recently, biblical stories functioned similarly in the contemporary 'West'. Folks could allude to a 'prodigal son' or a 'Good Samaritan' and be confident their audience understood the referent. In similar fashion, Bayle does not have to explain what the Augean stables are, nor the Hydra, not even a Chimera (a favoured word in his writing). The educational background of the seventeenth century is such that requisite reading and studying of antiquity's classics creates a common language around these stories, providing ample resources for similes. So Bayle can speak of the mythos of Jupiter saying the tales are "found in a great many books, which school-boys have daily in their hands" (*DHC*, Jupiter). An education in the classical mythos is assumed of the *Dictionary's* readership.

Roman gods were everywhere. Everywhere, for there was a nigh-infinite amount of divine beings of varying status and power. There could always be more gods unknown to humanity and new cults would pop up around worship of a previously ignored being. Yet any one individual god could be of little importance. There was always another. It was not always the Olympian pantheon that reigned supreme outside of the popular tales. A cult or town may prefer to worship a little known god but one that reflects them better or meets their personal needs. The mythos and pantheon was shaped by their literature. As Martin Goodman puts it, "For Romans, the *existence* of the gods was accepted by almost everyone as true but tales *about* the gods were freely described as the inventions of the poets" (2008, 264). This was a spiritually saturated world but what could be 'known' was wrapped in an admitted artistic creation that Christianity would reject in a search for, a need for, certainty.

Bayle accepts this poetic construction of the Roman religion and discusses it frequently (e.g. *DHC*, Chryssipus; Jupiter, N). This was the description provided by Cicero for the tales of the Roman pantheon. Does a poetic rendering make a tale 'untrue'? The Ciceronian historians acknowledging this construction seem to maintain adherence to a more generalised and sterile understanding of the deities of their world. A complete removal of divine presence is not implied. Yet it remains hard to escape the spectre of Bayle's sceptical doubt hanging over the many descriptions of religious stories as human constructions. Coupled with the constant parallels between failed flawed humans of all denominational stripes, we are left with a nagging itch that accepted Christian dogma may succumb to some of the same critiques as the attested falsity of Roman theology.

Goodman points out that "it is not easy to live with such a shifting, uncertain cosmology" and continues that Romans could often seek greater certainty in the regularity found in astronomy (2008, 265). Here is a great collision with Bayle's writing. Not only is Bayle at pains to point out that astrological considerations tell us nothing about the divine, the shifting uncertain cosmology is part and parcel of what it means to consider divinity. Certainty is fleeting, and often dangerous.

The worship of the deified emperors of Rome, post-Caesar, stands out as a notable distinction between Greek and Roman religion. These god-emperors also serve as a common talking point when discussions of virtuosity emerge in Bayle's writing. The awful excesseses of hubris and hedonism displayed from Roman emperors seem to carry an implicit challenge to divinity, at least to the seventeenth century classicist. The issue is not so much 'the emperors were obviously not divine beings' but more so 'how could this facade be maintained in light of such moral failings'—and the obviously human behaviour demonstrated therein.

The presence of the great classical writers permeates the *Dictionary*. Hardly an article can pass without a citation or reference towards Cicero, Ovid, Livy, Virgil among others. Often they are authorities. When they are not, they require reference in order to correct and amend. These sources remain central to the mental library of Bayle's contemporaries.[2] Even among the most infamous articles of Bayle, say 'Manicheans', references to Ovid and Cicero may be found. Antiquity is a constant referent for Bayle even when, strictly speaking, it is not necessary to do so. The classical world is where he draws many anecdotes and examples.

At times we see Bayle maintain the view that Rome was 'serious' and Greece was 'inventive'.[3] Thus Rome is a model for civil structure, martial prowess, and prose but the Grecian world is muse for creative philosophy and poetic imaginations.[4] Generally though, the two mould together in Bayle's retellings of shared religious pantheons with significant distinctions emerging in the deified emperors of Rome. One difficulty that will remain comes from *how* Bayle is using his sources. It's often easy enough to point out new information we have gained about classical sources that change the way they are interpreted or used from what Bayle had access to. This does little to address whether Bayle's approaches are 'genuine' or if he's taking some more liberties with classical texts in order to predate his own contemporary ideas.

A cursory glance through the list of articles in Bayle's *Dictionary* reveals name after name from classical antiquity. Scholars have inquired as to what topics make the cut for inclusion and what gets left out and the eclectic nature of Bayle's selection makes this a perennially intriguing question. In some cases Bayle provides an explanation. Biblical characters are alphabetically frontloaded (Adam, Abraham, David, Eve) and cross-referenced from those scant few articles, but were largely abandoned due to the saturated market for Bible references (*DHC*, Preface to the first French edition). That Bayle chooses to muse on so many classical figures indicates a personal interest in Greco-Roman culture and myths, though an unsurprising one.

It is a wonder we have as many articles centred around Greco-Roman culture as we do. Bayle embarked on the process of constructing his *Historical and Critical Dictionary* by first publishing the *Projet et Fragments d'un Dictionnaire Critique* in

---

[2] For an example of humanist debate over imitation of classical writers see Salmon 1987, 27–53.
[3] Classically expressed in Gay 1966, 94–98.
[4] For example, on Roman seriousness: *DHC*, Castritius "he maintained the grave Spirit of ancient Rome"; *DHC*, Claudius, A "Valerius Maximus observes, as a Proof of the ancient Gravity of the Romans, that the Magistrates always answered the Greeks in Latin, obliging them to make use of an Interpreter".

1692. The initial goal of the *Dictionary* endeavour was to function as corrective articles to the errors Bayle found within Louis Moréri's *Grand Dictionnaire Historique* (1674). The experimental *Projet* was designed to gauge reader interest in Bayle's encyclopaedic undertaking. Would people purchase an encyclopaedia to correct one they already owned? Some of the Roman articles were already written up for the *Projet*, but the reception was not what Bayle anticipated. Alas, a market is needed. In the preface for the first edition of the *Dictionary* we hear Bayle explain how the project had to be reoriented:

> Moreri is much more faulty in what concerns Mythology, and the Roman families than in modern history, I had particularly made collections upon the heathen gods and heroes, and the great men of ancient Rome. The work I proposed to publish, would have contained abundance of articles, like those of Achilles, Balbus, and Crassus, in my Project. All these vast collections are become useless, because I was informed that these subjects pleased few people, and that a volume in folio, the greatest part of which should run only upon such things, would be left to grow mouldy in the booksellers ware houses (*DHC*, Preface).

Unfortunately for Bayle, the market dictates his content and he informs readers that, regarding Mythology and ancient Rome, "Few articles of this kind will be found in my two Volumes; nor would they perhaps have been there, had they not been wholly composed before I was fully informed of the taste of the readers" (*DHC*, Preface). What this does indicate is perhaps a closer look is required when an article does make it past Bayle's self-sanction, what purpose is it serving if not correcting errors?

6.1 The Virtues of a Model Society

Much focus is placed on Rome as a model of the seventeenth century Republic of Letters. This self-conscious modelling sees members of the Republic of Letters (self-identified members of course!) compare themselves to the Roman poets and orators in an implicit nod to the 'civilising' behaviour displayed by the Ancients. It is hardly necessary to critique whether an enlightened state of affairs actually obtained in ancient Rome for—regardless of the real conduct of antiquity—Rome was idyllic and men of letters aspired to be a Seneca or Pliny born anew. Yet to endorse Greco-Roman ideas and character meant a tacit approval of paganism. Something *good* has come from this. Ah! But this gives vindication to some of Bayle's most infamous claims that moral virtue can be found outside of the Christian faith.

Bayle's oft-cited description of the Republic of Letters—"This Common-Wealth is a state extremely free."—occurs in the article 'Catius' (*DHC*, Catius, A).[5] A state extremely free, yes, but with a code of conduct gatekeeping entry. Even initial access to the circle of the learned was not truly open to the everyman, even as it

---

[5] It's worth mentioning that 'Catius' the historical figure is not directly the prompt for Bayle's comments on the contemporary Republic of Letters but rather a historiographical debate about whether this is the same person described by both Cicero and Horace.

circumvented traditional aristocratic privilege. The Republic of Letters opened up a state extremely free for the educated middle-class. For those who could access the conversations of the learned (Brockliss 2013, 71–100; esp. 87).

It is intriguing that Bayle's longer description here of the Republic of Letters sets up an image of 'innocent War' being waged. That anyone is so free they can wield truth and reason against any other. Importantly for Bayle, this is a 'just' war that can be waged by men of virtue against each other for the sake of figuring out knowledge when one or another gets it wrong. However this is, and Bayle is at pains to state, vastly different from satirists who defame without the truthseeking component.[6] For the virtuous wielder of truth and knowledge does so in order to pursue the truth, not to take down their opponent.[7] The concern with reputation is particularly pronounced when Bayle places it within an article on a man who died more than a millennium before. For the goal of a man of letters includes the longevity associated with scholarly virtue. That Catius is worth mentioning in Bayle's Dictionary is because his scholarship and ideas have long outlived his body. And for an educated class schooled in the rhetoric of classical Rome this was the model to emulate.

The article Accius (Lucius) sees Bayle use discussion of this Roman poet to enter the ongoing debates between the Ancients and the Moderns.[8] We are told, "some authors have expos'd themselves to ridicule for having imitated, or admir'd the Latin of this Accius, in an Age, when the Language was in much greater Purity" (*DHC*, Accius). The specific concern here is to what degree we should emulate the style of Ancient writers we admire? The problem with straightforwardly imitating is twofold. First, we run the risk of assuming the work we imitate is a good model to follow, when perhaps we are captured by the beauty of nostalgia. Second, we can adopt perfectly good style, while still failing to recognize how dated and hard to comprehend the language has become. Bayle reckons that "it is only in Latin, that Authors are yet found, who are fond of the oldest Phrases." Working to imitate fifteenth century French in the seventeenth century, one can grasp that the language has become 'superannuated'. We may still choose to write in that style, but would do so consciously to imitate, say, Marot. Latin runs into a peculiar problem. The people of letters understand it, but the language remains imitating ancient Romans when Latin came "to it's Perfection" (*DHC*, Accius, P).

Bayle's article on 'Brutus (Marc Junius)', murderer of Caesar offers some extended reflections on the virtues of the pagans, and by extension our own. Bayle describes Brutus as,

> so full of the great and noble Ideas of Liberty, and that Love of his Country. .. that neither the Obligations he had to Julius Caesar, nor the certain Prospect he had of aggrandizing himself as much as he pleased under that new Master of Rome, could balance the Ambition

---

[6] Bayle directs readers to his dissertation on libel found in the Appendix of the Dictionary.

[7] Bayle cites by name several men of letters of previous generations: Scaliger, Vossius etc. who emulate this 'no holds barred' truth seeking.

[8] See Helena Taylor's *état présent* in "The Quarrel of the Ancients and Moderns." 2020, 605–620. Taylor has also written a helpful piece about Bayle's historiographical approach in 'Ovid'; see Taylor 2017, 141–162.

he had of restoring things to their first State by the Murther [sic] of the Tyrant (*DHC*, Brutus (Marcus Junius)).

Immediately we can note that Bayle has placed a moral evaluation of Brutus and his dispositions in the main text of the article, usually reserved for 'historical facts', as opposed to the remarks where he tends to expand on his own opinions and thoughts. Bayle extolled the virtues of Brutus while concluding that it is a 'pity he sullied' these qualities by the murder of Caesar. The desire for liberty and fight against tyranny should not have been reduced to murder. Perhaps by following these virtues to their proper end, the intellectual battle, Brutus could have remained unsullied as a virtuous hero of Rome. The article on Caesar himself shows a similar praise and condemnation for the various aspects of Caesar's character. His military prowess and ability to lead can be praised but the death and suffering he inflicted must be condemned (*DHC*, Caesar).

Bayle's relating of Caesar's approach to religion carries forward this parallel with Roman virtue and contemporary Christianity. Bayle tells us the story of Caesar's siege of Marseilles where he ordered his men to cut down a wood belonging to a local deity. Unwilling to obey their commander, Caesar swings the axe first to convince his men that no divine wrath would befall them. Further, Caesar pledges that any forthcoming divine fallout would be upon him and not his soldiers for this potential blasphemy. But far from being an atheist, Caesar did hold various religious practices and beliefs. The gods really were present and active for the Roman commander. But this is, Bayle tells us, no different from most Christians today: "Christians do the like with the Directors of Consciences; they consult them, but do not obey them"(*DHC*, Caesar, H). If anything this is a strike against the position that Roman virtue can be emulated and modelled, especially when coupled with the heinous actions of war undertaken by Caesar. Yet, in typical Baylean fashion, this musing serves more to 'bring down' the religious disputes of Bayle's day than it does any resituating of Caesar's own dispositions. If we want to focus on the incongruity between Caesar's attested faith and his impious actions, we would do well to note how very commonplace this is across religious persons and condemn Caesar for his more obvious failings rather than the practical religion he held.[9]

Contra the virtuous positioning of Roman notables we find Bayle's condemnation of the pagan gods. The article "Jupiter" states bluntly: "nothing can be more monstrous than the religion of the Pagans, who looked upon such a God, as the supreme governour of the world, and suited to that notion, the religious worship that was paid him" (*DHC*, Jupiter). Once again though, the pagan religion is compared with Bayle's experience of Christianity. Whatever the gods of Rome *seem* to permit, and it's worth pointing out that divine behaviour need not be exemplary here, the Romans maintained a high standard of virtue.

Remark D explains,

---

[9] A similar discussion can be found in *DHC*, Cassius Longinus (Caius), I and K. Detailing one of the murderers of Caesar and the balancing between one's principled rejection of the supernatural and willingness to believe presages and prophecies in times of distress.

> It must be said to the glory of the Heathens, that they did not live according to their principles. It is true, that the corruption of manners was very great among the Heathens; but many of them did not follow the example of their false gods, and preferred the ideas of virtue before so great an authority. What is strange is, that the Christians, whose system is so pure, come but a little short of the Heathens in their vices. It is a mistake to believe, that the morals of a religion answer the doctrines of the articles of faith (*DHC*, Jupiter, D).

Morals seem to have little connection with the doctrines of the associated religion. Doctrines can point towards virtue or vice but humanity seems to maintain a moral status quo. Good and evil are found throughout diverse religious societies.

We see the same argument come to the fore of 'Lucretia' Remark E where Bayle returns to the question of how Roman virtue comes to be when the divine examples are so immodest. Here Bayle is pulling directly from his earlier work, the *Pensées Diverses*. The payoff is an emphasis on the natural ability to determine virtue, no gods necessary.

Likewise, the article 'Juno' has a fascinating exploration, contained in the main text, where Bayle wonders about the misery of a divine being. Bayle wonders,

> whether any of the wise men among the heathens took notice of a thing, which methinks was not difficult to observe, which is, that no body did partake less of a happy life, a state most essential to the divine nature, than the greatest of the goddesses (*DHC*, Juno).

There is a genuineness here to speculation about a supernatural being which aids the reader in reflection on the questions raised. This is not a mockery of the 'wise men among the heathens' but a concern about the limitations of their divine concepts and, in turn, what this says about humanity. The assumption in play is that the divine state must be 'la vie heureuse'. That might run against classical theology and divine impassibility but we could charitably understand this as 'satisfaction' and more importantly that the divine is not miserable, as Juno is described. The 'heathens should have made' these reflections and Bayle attributes a blameworthiness to the Romans' lack of divine understanding. These gods are too human.

> Yet it is the Philosophers who suffer when they inherit such convoluted dogma:
> The Philosophers who undertook to answer the Christian doctors, were very much to be pitied. They bore the punishment of the folly of others: the ancient priests had committed the fault, by ridiculously transplanting the fictions of the Poets into the public worship; and the Philosophers many ages after were forced to bear the shame of all these sopperies, and to torment themselves to put by the thrusts, that pierced them through (*DHC*, Juno, AA).

Remark DD makes explicit the connection with fantastical pagan stories and Bayle's concern with Christian theology saying, "there are many natural properties, which the traditions of the Christian people attribute to miraculous causes, as the Heathens attributed to Juno the fragrancy of the fountain" (*DHC*, Juno, DD). All people, ancient and modern, have a tendency to find the miraculous where they do not understand and must be 'undeceived' as they grow up. Bayle believes this is a benefit of a religiously tolerant state.[10] The diversity of views forces better defending of

---

[10] See also *DHC*, Launoi, Q "Some doctors of quicker fight, and greater courage than their brethren, will undeceive abundance of private persons, and yet cause no change in the public ceremo-

what one actually believes and can stop poetic fictions from becoming engrained only to later cause their defenders to stumble over the ridiculous.

6.2 Sovereignty and Providence

If we look at one of the most infamous articles of the Dictionary, 'Manicheens', we see the Roman gods feature heavily in Bayle's musings on theodicy. 'Manicheens', along with 'Paulicians', was the source of great controversy in Bayle's lifetime and beyond. In these articles our philosopher grapples with the insolubility of the 'problem of evil', that a good god could permit suffering in the world.

In Manichees C Bayle's interlocutor is Plutarch who Bayle quotes as saying, "the nature of God does not permit him to do any thing but what is good, nor to be angry with any one, nor to hurt him." Bayle's following gloss on this is, if not a little disingenuous, reflective of where the article's sympathies lie:

> this author therefore must have been persuaded, that the afflictions, which so often torment men, proceed from another cause than God, and consequently that there were two principles, one that did nothing but what was good, and the other which did nothing but what was evil (*DHC*, Manichees, C).

These sentences might seem straightforward and do not bring Christianity into the crosshairs explicitly, but Bayle's description of Plutarch's theology could map onto Christian dogma likewise.[11] Readers agreeing that Plutarch has implied a second divine principle exists, one which evil originates from, have found themselves stumbling into Bayle's trap. These are the same difficulties Christian theology hits. The remainder of the remark spends many words talking about the various pagan groups that, like Plutarch must have, subscribe to a two principle theory in their explanation of evil in the world. The nagging thought, inescapable to Bayle's Christian audience, is if they too 'must have been persuaded' to endorse such a theory in their defence of divine benevolence.

The great Athenian statesman Pericles makes up an important article for this study. In a similar vein to Bayle's articles on Pascal and Melancthon, there is an air of admiration around the Pericles discussion. This is a man who did not discard religion, but sought to make it rational and pare away the abundant superstition from Grecian worship. Bayle attributes this to the training of Anaxagoras the philosopher who "inspired him [Pericles] with a more rational religion, and he was not disturbed with superstitious fears, but he expected heavenly favours with a quiet mind" (*DHC*, Pericles, A).

---

nies. The ritual will last longer than the faith on which it was built." Bayle acknowledges that the 'argument from antiquity' is problematic. Just because something has been attested for a long time, does not make it true.

[11] God's anger or capacity to hurt may be debated if defenders located these under the divine attribute of 'goodness'. That God is benevolent was indisputable.

A long reflection on the 'doctrine of presages' follows in which the birth of a one horned ram was divined to signify that the split factions of Athens would consolidate under Pericles leadership. Anaxagoras, dissecting the creature, explains why the creature has one horn by means of biology by noting where the skull will grow. Recording these conflicting readings of the prodigious ram, Plutarch suggests that both the diviner and the philosopher could be correct: one explains the effect the other the cause.

The strongest reason to oppose this rational approach of the philosopher, Bayle reasons, is that causes and rules don't delimit what something may signify. The example provided is lighthouses, which signify where a navigator can safely travel but not necessarily so by how they are caused or generated. However, Bayle rejects this inference by saying, unlike lighthouses, effects of nature "cannot be presages of future contingents, unless they be appointed for that end by a particular intelligent being." The resulting conclusion is that meteors, floods, and one-horned rams cannot be miraculous productions unless one accepts that God produces "miraculously, and by a particular will all the natural effects which are looked upon as prognostics." The most puzzling element of Bayle's conclusion is how it seems to hinge on how we define and understand 'miraculous'. To hold with Plutarch that both the diviner and the philosopher may be correct in interpreting an event is, Bayle insists, to hold many natural effects as formed by God's particular will. Yet this seems a perfectly acceptable conclusion if one allows for divine particular will in creation. The 'prodigious absurdity' for Bayle is to assume miracles are "almost as frequent as natural effects." He suggests that a stronger prodigy could be indicated by God through the appearance of a one-horned ram that cannot be accounted for by the philosopher's examination. Fair enough. But this scenario can be pressured in both directions: why a ram at all and not some completely divine messenger? Or, why can the coupling of a natural cause not be part and parcel of the intended prognostic? The interpreter of Bayle will immediately associate this long examination of prodigies with the *Pensées Diverses*, Bayle's extended takedown of supernatural signs. Yet the earlier work is not referenced throughout the passage, even when meteors are mentioned as prodigious occurrences. Bayle is not shy to throw citations towards his own work, though often done anonymously. What should we make of the absence of the *Pensées Diverses*?

Perhaps we can read the Pericles article with an eye to Bayle's contemporary situation. Remark C is prompted by the statement, "They were so unjust as to suspect him of Atheism, because he had throughout learned the doctrine of the Philosopher [Anaxagoras]" (*DHC*, Pericles). The remark itself offers little to unpack this, a couple quotations from Marcellinus and Diodorus Siculus. Bayle provides us a gloss on the Diodorus citation saying, "They esteem a great man, when they are involved in a great war; but the sweetness of peace plunging them into idleness, they discover their jealousies, and prosecute him as a criminal" (*DHC*, Pericles, C).[12] It

---

[12] Cf. Pericles H "Which side can a reader take in the midst of so many malicious stories?"

is hard not to wonder whether Bayle had himself in mind in this framing of Pericles.[13] Did the safety of the exilic community in the Dutch Republic, a sweetness of peace, encourage Bayle's Huguenot church community to 'prosecute him' for impieties?

Plutarch offers a reflection on the character of Pericles which Bayle picks up to muse on divine nature. Plutarch felt the nickname of Pericles, Olympian, was well suited due to the kindness and integrity of Pericles in positions of power. This, to Plutarch, fits with the notions of gods as authors of all good things and not evil. Contrarily the 'Poets' picture the gods as 'full of trouble, and enmity, and anger, and other passions' (*DHC*, Pericles, K). Here Bayle will spring into action and connect his exposition with his most infamous articles 'Manichees' and 'Paulicians' for Plutarch has committed the Manichean fault! By only allowing G/gods to be authors of all good things, Plutarch has defended a two principled approach, smuggling an evil principle into the world from the negative space of his statement. And it is not only 'Manichees' and 'Paulicians' which intersect in this Baylean web but the earlier article 'Jupiter' as well, as Bayle interrogates common conceptions of the leader of the Greco-Roman pantheon. Though often known as divine through 'thunder and lightning' Bayle, citing Cicero for support, defends a general notion that "nothing is more agreeable to the divine nature than to do good" and Jupiter was firstly a 'helping father' (*DHC*, Pericles, K). Several passages follow in which 'godliness' or becoming like/a god (apotheosis) is associated with goodness albeit sometimes the goodness of creating a useful thing, or aiding the flourishing of humanity.

Nonetheless, Bayle contends that "Sound theology agrees with all these notions of the ancient Heathens." God is "infinitely more inclined to use mercy than severity"; when we do good we act in godlike fashion (*DHC*, Pericles, K). Wherein lies the condemnation of Plutarch? Well, allegedly Indigenous Americans could view the invading Spaniards as gods due to their firearms. This "shows that two opposite things lead men to the knowledge of God; *viz.* The power to do ill, which he exercises so severely, and the goodness wherewith he bestows a thousand benefits upon mankind" (*DHC*, Pericles, K). Debating with his own earlier article from the *Nouvelles de la république des lettres*, Bayle finds the ills/goodness calculus difficult to tip in the balance of goodness (NRL: OD I, 110). We can find a thousand benefits God bestows, sure, but ill strikes us so much harder than everyday pleasures. We remember the misery of "fifteen days sickness" as much as "fifteen months health". This is a cynical outlook, absolutely, and should keep our gaze on the aforementioned links to the 'Manichees' insolubility of evil front and centre in our reading. Whatever we make of divine presence, Bayle will not let us shake off the world's ills whether we personally have faced them or not. An account must be given:

---

[13] The more praise-worthy subjects of the Dictionary are often probed for authorial likeness. See Tinsley 2001, 88, 96.

> The mind of man, being too narrow to comprehend clearly, how the miseries and crimes which cover the whole earth, can be consistent with an infinitely good Being, has run itself into the hypothesis of two principles (*DHC*, Pericles, K).[14]

One final thought worth mentioning on this remark is the paralleling of Jupiter with 'God' throughout. On the one hand, it makes much sense to render ancient notions of divinity with the lens of the Christian God when working through these larger debates. Yet on the other hand, this is also creating a juxtaposition many Christians would want to deny: that Jupiter is in any way like the Christian God. The classical theistic attributes simply do not map onto the deities of the Greco-Roman pantheon. These are beings which lack the 'omni' qualities monotheistic religions ascribe to God. Yet Plutarch making the movement to a more 'rational' god in his rendering of Jupiter—combined with the parallel of Jupiter to 'God' throughout—brings a sceptical flavour to the attaching of 'omni' qualities to the God of Christian scripture and tradition. Does benevolence really fit here? What about omnipotence? Are these at odds with the narrative of Scripture? Have we 'rationalised' God to create a being worthy of worship? These questions emerge throughout Bayle's articles, and the carefully crafted links between various sceptical framings of evil and divinity should remain highlighted.

As we can expect of Bayle as soon as these questions are implied, they are pulled back and hedged. The praise of Herodotus's understanding of God will be the site of this battle. Reciting Henri Estienne's description of Herodotus, Bayle records "this opinion of Herodotus about the power and providence of God, is the same which Christians ought to have: he pretends that it is impossible to say any thing more divine upon this subject" (*DHC*, Pericles, L). By virtue of his knowledge of God, Herodotus is deemed to possess a piety 'as great as it could be in a man destitute of the light of the gospel'. This is well in line with speak of the 'enlightened ancients' that they may possess a piety, but lack the gospel, or true knowledge of God. But Bayle cannot let this praise go unchallenged. He points out that Herodotus still speaks of 'jealousy' in God. Holy Scripture ascribes many things to God which we read as metaphorical, "eye, hands, anger, repenting, joy and fear," but, for Bayle, Herodotus has crossed a line. The historian's understanding of divine jealousy is not the Christian description of divine sovereignty, a jealousy against idols and other objects of worship. Rather it is an anthropomorphized jealousy of human goodness and success. In a comedic turn of phrase Bayle says God does not delight in disrupting human success "in his hours of recreation, and that it is his tennis" (*DHC*, Pericles, L). Now of course, Bayle adds, this is not to dispute the 'vicissitudes' of God so often described. For God does enjoy raising the lowly and humbling the powerful, but this is quite different from the removal of blessings and punishing of success *because* God feels threatened by human exemplars. Whatever that form of jealousy is, it is not divine.

---

[14] Bayle's familiar solution is to fall back to Scriptural revelation, a fideistic cry that we cannot know how evil and the goodness of God interplay. Whether this solution is offered in good faith, or as an undermining of religion, remains a compelling debate for all engaging Bayle.

Talk of the two principles resumes in 'Xenophanes' prompted by a claim by this philosopher that there is more good than evil in nature. We can make this calculus, Bayle supposes: "In order to know whether moral good equals moral evil among men, we need only compare the victories of the Devil with those of Jesus Christ" (*DHC*, Xenophanes, E). Christ being the representative of men as mediator. The Devil, at least popularly conceived, indicates the success of the two principles. Bayle explains, "How detestable soever the opinion of the two principles hath constantly appeared to all Christians, they have nevertheless acknowledged a subaltern principle of moral evil" (*DHC*, Xenophanes, E). Ah, but alas, Bayle adds that the historical tally is firmly on the side of the Devil, with only a few triumphs of Jesus Christ noted. "Thus whilst the Devil reigned uninterrupted without the limits of Christendom, he so successfully disputed the ground in Christendom, that the progress of his arms was beyond comparison greater than that of truth and virtue." Perhaps the 'uninterrupted reign' of the Devil being not so different from within Christendom is to give us pause here. Bayle cites Jurieu's *Le vray système de l'Église* (1686–1689) for support that popery is a place where the devil thrives. But, Bayle quickly adds that Protestants are not spared from Jurieu's critique, and that debauchery exists across the Christian spectrum (*DHC*, Xenophanes, E). There will be more creatures who hate God than creatures who love God, based on the Devil's success, is this not a victory for evil?[15]

A reference to Remark I of the fabulist Aesop's (Ésope) article directs the reader to continued discussion on the humbling of the powerful with a quote attributed to Aesop that would be right at home in the Christian Bible "he lowers things that are high and raises those that are low"[16] (*DHC*, Esop, I). Biblically astute readers will recognize the commonalities, but will be rudely confronted by Aesop's quotee, the God Jupiter. The remark continues by playing with the image of a 'Wheel of Fortune'. Success and failure are largely random. Those that achieve greatness have spun the wheel well, but there is no guarantee that their descendants will continue to succeed, nor any necessary correlation between their actions and their fortune. The wheel simply spun well. Likewise for the world's failures. But the *Rota Fortunae* requires the hand of Fortune to spin, changing the worldly positions of those the goddess targets. The imagery demands we ask if life's fortune has a divine hand spinning the wheel. And if so, is the outcome random, or only random insofar as our fortune will change. There are also exceptions, say, the Roman empire succeeding for so long. And if the spinning wheel is a rule, Bayle questions why humanity cannot change:

> two thousand Years hence, if the World should last so long, the continual Revolutions of the Wheel will have made no changes in the Heart of Man. Why then are they repeated without Intermission or end? Here it becomes us to adore in humble Silence the Wisdom of the Governor of the Universe, acknowledging at the same time the infinite Corruption of our

---

[15] This remark carries links to Pericles K on sickness and suffering (small amounts of suffering counts for many days of health in balance) as well Manichees C. See van der Lugt 2016, 33–46 for an extensive look at the 'Manichéens' web.

[16] Cf. Matt. 23:12, Jm. 4:6.

> Nature and it's Servitude under the Yoke of machinal Impressions, an inveterate Disease which yields to nothing but the miraculous Operations of Grace (*DHC*, Esop, I).

A Reformed piety is expressed here but couched within the larger Baylean remark, and amidst the workings of pagan deities, it is difficult to parse what the author wants to stake as the human situation. Bayle imagines a moral pendulum, that should we possess the divine knowledge of the totality of human corruption and how grace operates, we could chart almost metronomically how the prideful will be humbled and the humble raised up according to Aesop's response.

In a similar manner, a reflection in 'Timoleon' allows Bayle to remark on the Corinthian general Timoleon's acknowledgment of the gods in his successes. Timoleon built a chapel to the goddess Fortuna, recognizing his victories as the "favour of heaven". Bayle thinks this architectural investment shows a genuineness in Timoleon's claim but notes how often divine providence is cited in bad faith to further political ends; "God was on our side". Apparently, the character of Timoleon, as one who actually accomplished great deeds, makes it unlikely that he would appeal to Fortuna in an attempt to puff up and hyperbolize his successes:

> Such a man as he could not be ignorant, that he diminished the merit of his prudence and valour, in proportion as he acknowledged that Fortune as the cause of his victories. How therefore could he make this acknowledgment out of a principle of vanity, supposing him to be guilty of a lie? (*DHC*, Timoleon, I).

The virtue aspect thus remains forefronted in discussion of pagan worship. This further undercuts certain positions on 'idolatry' by highlighting the genuineness Timoleon worshipped with, and his humbling posture towards the goddess. This is no willful delusion but a man displaying thoughtful and appropriate religion even within a 'false' religious system. Remark K allows space for Bayle to further discuss the role of Fortuna in pagan notions of divinity, following a recognition of the complex entangling of 'fortune' and 'providence' for the Christian. Here the 'Philosophers' who recognize the "unity of the deity" are recast in support of a generalised classical theism, recognizing the actions of Fortuna not as fanciful whims nor random chance but actions occurring with good, though perhaps misunderstood, reasons (*DHC*, Timoleon, K).

Locating 'Fortune' with God does not simplify or resolve any of the fortune/providence dilemma for the question remains whether actions, good and ill, happen directed by the divine will or by happenstance (and if happenstance is possible with any standard account of divinity). What this massive remark on fortune amounts to is another angle of a typically Baylean address to the insolubility of evil. Appeals to chance must still answer the role of divinity in their schema. Scripture can be brought into the conversation,

> we ought always to remember, that our Theology, and the universal language of all Christians, founded on all passages of holy writ, lay it down as a certain truth, that the blindness of man, his rashness, folly, and cowardice, are frequently the effect of a particular providence, which inflicts them on him as a punishment; and that his prudence, his wise answers to questions, his resolution, and his understanding are favours inspired by providence, in order to his preservation and prosperity (*DHC*, Timoleon, K).

But what might seem like an answer and explanation of providence runs against the gauntlet of experience. The question remains, does God bring bad Fortune?

Tullie Q relates Cicero building a temple to honour his deceased daughter Tullia. The remark deals with factual questions surrounding whether this temple was indeed constructed. However the most interesting component emerges when Bayle contrasts Cicero's own words against the deification of deceased Romans with his own apparent desire to see his daughter deified upon death. Contextually, Cicero is writing the *First Philippic* against Antony's attempted consolidation of power in the wake of Caesar's death. Thus the lone citation Bayle provides from Cicero against a deification of Caesar can be confidently understood as calculated Ciceronian rhetoric against a political opponent. Cicero will go on to advocate for the deification of Caesar at times when it supports his political agenda (Cole 2014, 1–8; 170–174). Bayle's point, though inaccurate with regards to Cicero, is that grief can cause us to muddle our principles. Cicero's willingness to "sacrifice the honour and glory of his gods to his ridiculous fancy of deifying his daughter… affords a natural image of the empire of the passions. They spare nothing in heaven or earth, when bent on their justification" (*DHC*, Tullie, Q). Most intriguing is the attached citation to this comment, a somewhat cryptic reference reading: "We have seen of late a famous minister, who endeavoured to find in the Prophets of the Old Testament, all the faults that were objected against the false Prophets of Dauphiné, whom he obliged to defend as true Prophets" (*DHC*, Tullie, Q). Those familiar with Bayle's conflicts will recognize this as one of his many cloaked references to Pierre Jurieu.[17] The attempt to find parallels between Old Testament Prophets and the Dauphinois prophets is likened by Bayle to Cicero's attempt to show Roman gods who were formerly men (chiefly Romulus) in order to support his push for his daughter's deification. Bayle's praise for Cicero's grieving process and the 'beauty' with which the rhetorician writes of loss seems at odds with the Jurieu insertion. We might opt to charitably read Bayle here as indicating how being stricken by grief can compromise consistent principles we may hold as we bow to the empire of the passions. This angle allows the critique to remain strong, 'don't justify false prophets', but shows an empathy from our philosopher: 'nevertheless, I understand why this parallel was made in times of exilic suffering'.

Cicero's claim that "men come into the world only to suffer the punishment of their sins" is rejected by Patristic apologist Lactantius for how it frames God's reason for creation. But Bayle is not so quick to dismiss the view, noting the parallels with orthodox Christianity. Bayle reasons that Scripture teaches all are "children of wrath" (Eph. 2:3) and the doctrine of original sin implies that all souls post-Adam contain sin and incur punishment for that sin. Whatever particular quibbles Lactantius has with Cicero's claim they must be better articulated and emphasise the many reasons, taught to us by revelation and not the natural light available to pagans, for God to create. It stays open to us readers whether we think Bayle's check on

---

[17] The reference would appear to be Jurieu's *Les Lettres pastorales aux fidèles qui gémissent sous la captivité de Babylone* (1686–1689), particularly Oct 1, 1688 re. Isabeau Vincent.

Cicero is genuine or if the balance of evidence that God permits punishment for sin in excess would lead us to dismiss the goodness and justice, or existence, of God.[18]

6.3 Scandal

We can examine a series of articles which play with scandalous material from Greco-Roman religion. The article 'Flora' relates to a courtesan, or perhaps a goddess, and the games and festivities around her celebration. Bayle has some of these spicy topics related by appealing to history to shield his own authorial selections. He is debating the historical subject of the festival, so discussion of why a courtesan should be celebrated is appropriate fare. Likewise, 'Etampes' discusses, at length, cuckolding and how this has been known from classical times as a way men allow themselves to get ahead by being cuckolded by royalty. Just doing the historian's job! Recording the scandalising facts of how people climb rank. 'Periander', the tyrant of Corinth, is "a very sensible proof of the disorder wherein false religions leave the hearts and minds of men. They do not correct mens inclinations to sin" (*DHC*, Periander, B).[19] Periander commits incest, necrophilia, and sheds innocent blood but is compelled to offer a golden statue in worship. Belief in gods does not correlate with moral uprightness. Many of Bayle's articles detailing a female subject are often dealing with classical figures, and often, sexual promiscuity.[20] The historian's responsibility in 'just reporting the facts' is used as a defence allowing Bayle to be while he includes titillating content in the Dictionary.

'Lupercalia' is an article which emerges despite Bayle's self-sanction on Roman topics. Further, it is one of the relatively infrequent Dictionary headings that is not a person. The Lupercalia festival was a celebration held on the fifteenth of February, led by priests known as Luperci, where women would be whipped with the hide of a sacrificed goat to promote fertility. It is an odd article for inclusion, it has few cross references or obvious connections with contemporary issues Bayle wishes to explore. There is a brief line in the main text where Bayle discusses the the strangeness of Christians supporting the persistence of this festival in late antiquity, "when at last, Pope Gelasus, in the year 496, would tolerate it no longer, there were Christians found amongst the Senators themselves who endeavoured to maintain it" (*DHC*, Lupercalia). Perhaps this speaks to greater concerns about traditions persisting because people become comfortable with ritual even when they should no better

---

[18] *DHC*, Tullie P, 412; 414 sees Bayle connect this remark to others in Xenophanes D and Pericles on the utilitarian calculus of good and evil.

[19] Bayle's descriptions of Christian moral failings make statements like this about 'false religions' suspect in both directions. Are all false religions because inclinations to sin remain, or do true religions still contain bad people?

[20] See also: 'Licinia', 'Lycoris', 'Lucrece', 'Penelope', 'Servilia', 'Thais' amongst others. 'Sforza (Francois)' contains a reference to Roman maids being inspected for imperfections at the temple before their marriage and connects to 'Fulvia', 'Hipparchia'.

as discussed in the 'Juno' article. It may be easier to conclude that the article made it past the 'ancient Rome pleasing few people' rule both because it keeps several corrections in place (Moreri and du Boulay's accounts) and because, of course, the festival involves fanatical whipping of naked bodies.

'Ganymede' consists of some of the most extensive discussion of same-sex relations in the Dictionary as Bayle discusses the rape of Ganymede by Jupiter.[21] The explicit connection to church judgments: "This justifies the fathers of the church, who have reproached the heathens with this villany of the greatest of their gods" and references to Adonis, Longus, and Chrysippus just in this remark alone convey a hotspot of controversy if one went looking for this content (*DHC*, Ganymede, B). Certainly, turning to 'Ganymede' would be a natural location for anyone interested in reading more scandalous thoughts on pederasty and could be a launching point to these other articles Bayle connects. Thinking about this sort of 'reading practice' is useful for reconstructing Bayle's own leaps and connections between the various articles of the Dictionary. What headings lead to what topics? Where does Bayle find himself connecting articles?[22] If you go looking for well-known scandalous stories from mythology, you will find them in the Dictionary.

'Tiresias' functions as yet another place for Bayle to write about lewd material. Retelling the story of Tiresias from Ovid he describes how Tiresias encountered two snakes mating. Upon striking the snakes Tiresias transforms into a woman, later turning back. Having been both man and woman the gods Jupiter and Juno call upon Tiresias to settle whether man or woman experiences the most pleasure during sex. When Tiresias sided against Juno, though in favour of women, the goddess blinded him. Now, on whether this judgment is supported Bayle says that as to natural reasons, you would have to ask a doctor. But as for moral reasons Bayle thinks Tiresias's judgement is correct for it is only fair that God would multiply the pleasure of women during 'that act' to make up for the "loathings, inconveniencies, and pain" women experience from conception to delivery! (*DHC*, Tiresias).

Deeper in the previously examined Pericles article, we read about a group of prostitutes who financed the construction of a temple to Venus (*DHC*, Pericles, T). They earned a great deal of money following Pericles' soldiers around on expedition but dedicated much of this to religion. Bayle takes this opportunity to compare the prostitutes to 'Financiers' who exploit the labour of others to grow rich and spend their money constructing chapels and altars. Though not a kind comparison for sex workers, the implication appears that religious affection is dishonest when

---

[21] If this was not enough scandalous content for Bayle's readership many remarks in 'Juno' and the 'Jupiter' article itself are also largely devoted to the god's sexual escapades.

[22] See van der Lugt 2016 on navigating 'webs' of interrelations in the Dictionary. See also Cotton 2020, 123–140 for an example of the navigation between articles connected by an enticing subject. 'Salmacis' is part of a web of articles where Bayle engages discussions on intersexed persons. This article relates the mythical fountain at Halicarnassus where the nymph Salmacis observed Hermaphroditus, son of Venus and Mercury, bathing. Being entranced with Hermaphroditus, the nymph pushed herself upon him and begged the gods to join them. The gods listened and joined the two persons in one intersexed body, a 'Hermaphrodite'.

engaged in moral impiety, and undertaken only to enhance or cover public perception of piety. Venus worship reappears in 'Sulpicia', a Roman lady who consecrated a statue to Venus 'converter of hearts' to reclaim women from their leudness into chastity. Sulpicia was chosen as she was considered the most chaste. Bayle notes what an honour this must be for who would acknowledge that someone else is more chaste than they are. But is Venus the correct choice for divine praise of this action? "Perhaps it will be said that the senate did not make a right application; for, according to the doctrine of the Heathens, Venus equally presided over unlawful and lawful love, and she had occasioned the prodigious leudness they intended to put a stop to" (*DHC*, Sulpicia, A). Bayle responds that those who cause great evils ought to end them, ergo, going to Venus with worship was the right decision by the senate.

'Thesmophories', as Bayle describes it, was a festival celebrating Ceres as lawgiver, though she was better known as goddess of the harvest. Only women were allowed to participate and during this time they did not lie with their husbands. Bayle, of course, latches onto the titillating part of the festival and expands at length about what enabled the women to 'suppress' their urges. Bayle is ultimately skeptical that the women would need to take any herbal substances to avoid temptation during the few days they are apart from their husbands. If this herb indeed suppressed, Bayle comically writes, "[it] would have been worshipped as the tutelar god of married men, and a *Deus averruncus* (God of harm aversion), or *alexicacus* (God of evil-aversion), with respect to cuckoldom" (*DHC*, Thesmophories, B). Despite his stated concern that this speculation itself calls into question the virtue of women of history, Bayle carries on his own exploration of just how much sexual abstinence a marriage can endure and how difficult this would be. A further remark on this festival remains where Bayle describes how the festival would feature a parade carrying a representation of female genitalia made of meal and honey. This was, surely, well suited to a celebration of Ceres for the good humour she was restored to, after losing her daughter, upon seeing an old woman dance naked and comedically in front of her. In all this concern around the celebration of womanly parts and how to remain chaste in doing so, Bayle believes the vigils and exclusion of men provide much more certainty of purity than any herbal remedies and thus falls into the same sapphic erasure as contemporary NASA on their alleged claim that all-women crews will avoid sex on the journey to Mars!

Bayle continues to make comedic remarks in 'Philyra' discussing the relations between this daughter of Oceanus and Saturn. First, on Saturn's equine transformation and flight from Philyra when his wife, Rhea, discovered their affair Bayle suggests that this is perfectly reasonable for what husband wouldn't run away if exposed to "hard words" and "scoldings". If only we could all be horses. His second remark gets even naughtier as he muses over whether Saturn transformed into a horse preemptively to better "enjoy Philyra" (*DHC*, Philyra, A). The racy parts of the Dictionary go on and on with throwaway references in many articles that need not centre such material. Two things stand out: First, the historical gap allows ancient or mythic figures involved in scandalous content to be safely described and recorded while maintaining the attractive appeal of such stories. Second, 'Gods' and religion are a safe place to discuss scandal. Yet even when comparisons with acceptable religion are not made, the issue of virtuous conduct confronts the reader of salacious material and questions of sincerity surround the maintenance of virtuous norms.

6.4 False Gods

'Philomelus' sees Bayle explicitly address the falsity of Greek deities in the context of the plundering of a temple to Apollo. He glosses the event as follows:

> For tho' the temple of Delphos was consecrated to a false god, yet it was an impious thing, and a sacrilege to plunder it, when they believed that Apollo was a true God. I own, that none but the true God can change the nature of prophane things, they cannot become sacred but by his institution (*DHC*, Philomelus, E).

There are plenty of occasions, some discussed above, where Bayle speaks of Greco-Roman deities in the mode of a storyteller, presuming the veracity of these beings within their narrative. So what stands out at this point, and why the story is addressed by the contemporary narrative seems to be the importance of conscience for the sacrilegious looters. It is not that they truly damaged the sacred, for their god and his ordained relics are falsities. Rather, that they *believe* Apollo to be a true God makes this an impious action. Blasphemy becomes a crime of the subject. More interesting, and congruent with Bayle's positions on other religions, blasphemy towards the Christian God cannot be a crime of non-Christians. You cannot blaspheme something you do not believe exists.

Bayle's article on 'Abaris' tells the story of Abaris the Hyperborean, a man granted an enchanted arrow by the god Apollo (*DHC*, Abaris).[23] By grasping the arrow Abaris can travel around the world. Bayle riffs off of this tale to bring up modern 'magic rods', particularly the dowsing rod of Jacques Aymar. A popular story to debate and discuss, Aymar claimed to be able to divine land boundaries, Springer-esque parental disputes, and even unidentified murderers. Of course, the Baylean scepticism applied to the tale places divining in disrepute and discusses the nature of the charlatan that undertakes such deceiving. But importantly for us, this all falls under the discussion of gods bestowing abilities. Are we meant to make a connection between the Aymarian charlatan and diviners of antiquity? What about the modern prophets and saints of Christendom? It is no secret that Bayle rails against prophecy and the dangers therein throughout his corpus. Bayle appeals to Leibniz here citing his take on the Aymar case: that we would be better spent asking how so many people could be deceived by Aymar than speculating on how such a divining rod could conceivably function, fake or not. People, with their moral failings and naïveté, are always central to our discussions of the divine.

The Greek philosopher 'Xenocrates' is considered to have "contemptible theology" for believing in eight deities, the seven known planets (including Sun and

---

[23] See Vermeir 2012 for an extensive look at this particular article including Bayle's construction and retooling of 'Abaris'.

Moon) and the plane of fixed stars. Bayle sees this as an opportune place to denounce viewing the planets as deities not only by Revelation but also by Nature. Perhaps contrary to Bayle's other discussions of 'gods' here he writes that from Nature alone we conceive of God as a singular being and so attributing deity to each planet is absurd.

> Writing on another philosopher 'Pythagoras' Bayle records how in preaching against the vanity of his culture, Pythagoras convinced youth to forgo jewellery and fine clothes, sacrificing these to the gods. Bayle here continues his comparison of ancient exemplars with contemporary Christianity. Once again, the ancients triumph. Not only on moral performance but as conversionary preachers exceeding the evangelists of the day for "These Christian preachers could not do what a heathen Philosopher did" (*DHC*, Pythagoras, C).

The praise for Pythagoras does not stop there. Like Pericles, Pythagoras is viewed as approaching orthodoxy, "The Heathen Philosophers never said any thing finer, than what he said concerning God, and the end of all our actions; and it is likely that he had carried his Orthodoxy much further, had he had courage enough to expose himself to martyrdom." What does this mean? Bayle expands upon this in the attached remark but at face value it sounds bizarre. Surely 'orthodoxy', traditionally conceived, requires an acceptance of the Christian creeds, which an (pre-Christian) ancient philosopher cannot do. Orthodoxy is framed here in the philosophical mode, separating 'God' and beliefs about the rational god of philosophy, from the meat and bones of Christian practice. Pythagoras is related to have "acknowledged the unity of God" but what this actually entails is frustratingly vague, both for the ancient sources and Bayle's summary (*DHC*, Pythagoras, O).[24] Josephus is cited as believing Pythagoras could have spoken "more soundly still" if he was not afraid of persecution from the dominant religion thus the aforementioned fear of martyrdom. The Pythagorean unity may approach a monotheistic conception (though Pythagoras' own involvement in traditional religion remains unclear), but it is quite different from a Christian conception of monotheism. What is being achieved in relating Pythagorean dogma in this praiseworthy mode?

The expanded remark dials this back somewhat, pointing out the unorthodox elements of Pythagorean dogma: two independent principles (the Manichaean dilemma again!), a pantheistic incorporation of souls into Godself. The more interesting question, it seems to me, is why try to frame Pythagoras, among others, as 'orthodox' in any meaningful way. The obvious response seems to be the desire to endorse various elements of Greco-Roman philosophy. In showing the alignment of, say, Pythagorean thought with Christianity, one illustrates the shared virtues that develop from similar beliefs. But this structure upholds a virtue-construction framework that seems unBaylean. Bayle often seeks to divorce Christian beliefs from moral living, in large part simply by illustrating the failures of Christians. This approach is further highlighted, as we have discussed, by showing how pagan virtue can be *more* moral than many Christians. Perhaps here is a clue that Bayle might have something else in mind in positioning pagan beliefs as not only virtue forming (or maybe better

---

[24] De Smet and Daniel 2021, 288.

'virtue-permitting'), but as near-orthodox. But the praise continues to build after challenging Pythagorean 'orthodoxy' for Bayle suggests that the Philosopher is positively Christian saying: "Nothing can be more admirable, nor more Christian, than what Pythagoras said... for he taught that the study of Philosophy should tend to make men like God." This is expounded further saying to do the works of God, for Pythagoras, are to speak truth and do good. What could be disagreeable about this? Studying philosophy should make us speak truth and do good. In locating 'good philosophers' and remarking on them in the *Dictionary,* Bayle shows an alternative to orthodox piety.[25] If we know orthodox Christians commit some of the most heinous crimes, and we want to maintain that right beliefs beget right actions, maybe the category of orthodox has to be challenged. Positioning heterodox dogma alongside established belief systems challenges the power that orthodox systems can wield to nefarious ends.

The 'Philosophes Païens' are praised in 'Plotin' as well. Bayle relates of Plotinus that "he died as nobly as a Heathen Philosopher could do" saying, "I use my utmost endeavours to return that which is Divine in me, to that which is so in the universe" (*DHC*, Plotinus). Bayle sees in Plotinus's spirituality the 'seeds of Quietism' and a mysticism not unfamiliar to Christianity (*DHC*, Plotinus, K). That a unity with divinity can be achieved. That said, this is the 'god of the philosophers' which Bayle sees as approaching Christianity. Not a personal divinity, certainly not the identifying of the divine with a Christlike figure. It's intriguing to say the least that Bayle finds the positive virtues displayed in those who have a more abstract notion of the divine relative to both the pagan gods and orthodox Christian descriptions.

Describing a sophist who wrote on divine providence, 'Theon' argues that being persuaded that the gods see all we do helps people live in safety and practice duty, we have more pleasure when we believe the gods watch over us. Bayle agrees that *if* "men knew how to live according to their principles, the doctrine of God's presence would be the most effectual means to divert them from all evil, and to make them pursue good" (*DHC*, Theon, A). Of course the implication is that we do not live by our principles and such doctrines are ineffective. The pairing with a contemporary account of the Marshal de Gassion musing that had he really believed the doctrine of the Real Presence he would spend his life prostrate at church indicates the sort of principles Bayle has in view. Less that we do not live according to our moral principles but more so we do not live according to doctrinal principles, the invisible assertions of faith that the religious claim to adhere to, hold very little impact on their day to day lives. 'Theon' continues by asserting that apart from a few philosophers the pagans believed:

> That the Divine Nature was a sort of being divided into several individuals. They attributed to every god a large share of power; but did not exempt them from the imperfections of our nature; they believed them susceptible of anger and jealousy, literally speaking; they made no scruple to affirm, in their most serious writings, that a malignant and secret envy of the gods opposed their prosperity (*DHC*, Theon, K).

---

[25] La Mothe le Vayer's *de la Vertu des Païens* (1642) stands behind much of this discourse and is cited at times by Bayle.

What the few philosophers get 'right' in making the Divine Nature 'big', they paradoxically lose the explanatory potential of anthropomorphic, meddling, deities offer for dealing with human prosperity.

A different take on the 'false gods' angle comes in 'Scamander'. The article discusses the etymology of the river Scamander and the relation of the river to the river god of the same name. In a disturbing account Bayle shares the story of betrothed maidens who would ritually offer their virginity to the river. Lying in wait, was a man who claimed to be the river god and lay with the maiden. Even in an age of 'wit and learning' the "fatal power of a false religion" abounds. While the victim blaming from Bayle is cringeworthy, the seriousness he grants to pagan beliefs is notable. These are not stupid people. Anyone of us, in similar contexts, could hold the same beliefs and fall to the same ruse. Speaking of the divinity and worship of Romulus, we read "If other things than learning and science had not interposed, the divine worship of Alexander, Caesar, Augustus, etc. had continued as long as that of Hercules and Romulus" (*DHC*, Scamander, D). At the height of Greco-Roman science and culture we still see worship of leaders exalted to divine status. This 'idolatry' is not something that can be reasoned away. Presumably, the 'other things' than learning and science Bayle has in mind are the coming of Christianity to the Roman empire, though it is unclear.

Lest this stark dichotomy between pagan and Christian virtue seem uncharacteristic of Bayle, he inserts a musing that monks may try a similar ploy invoking a saint but the nature of purity and saintliness in Christianity would make it less likely to succeed. Presumably the meagre defence is that Christians of similar devotion would not believe a saint would wish to lie with them and would refuse the supernatural proposition. Importantly, the contrast between Christian and pagan here becomes happenstance for in another scenario the 'monks' could be just as devious and the Christian duped as well as the pagan maid. Both religious systems can be exploited by evil people. When Bayle concludes this segment saying: "There is nothing so advantageous to man, if we consider either the mind or the heart, as to know God well: On the other hand, nothing is so fatal to all the faculties of a reasonable mind, as to know God ill, as the Pagans did", we must wonder which religions really know God well and which know God ill (*DHC*, Scamander, D).

6.5 The Challenge of Pagan Virtue

'Marcellin' (Ammien) continues the prevalent theme of the virtuous pagan. Bayle asks why some have attempted to retroactively claim the historian Marcellinus as a Christian. Is this really so much better than admitting a virtuous pagan? Bayle provides his evidence:

> would a Christian, who wrote his history under the emperors who reduced Paganism to the last gasp, have been contented with speaking honourably of the Christian religion, and have never carried the matter so far, as to declare sometimes, that it was the only good and true religion, and that the worship of the Pagan deities was idolatry? Under such emperors,

would a Christian have been so inconsiderate, as to praise Julian the Apostate to the skies, and never declaim sharply against his apostasy, and his hatred of Jesus Christ? Would he have spoken of Mercury, and the goddess Nemesis, and the goddess Themis, and the augural superstitions of Paganism, as Ammianus Marcellinus speaks of them? I know of no Christian authors who did not, even in the hottest persecutions, speak contemptibly of the Pagan idolatry, and in some manner insult over it; and it is incomparably more easy to conceive, that a Pagan should use moderation, speaking of the Gospel, than that a Christian should do it, when he speaks of the worship of false gods (*DHC*, Marcellinus (Ammianus), B).

The concluding sentence of this assault is a dagger to the heart of refined Christian piety, that a pagan could speak in a more tolerable form of his religious dissenters than a Christian. Still all hope is not lost for the Christian faith within this article, as Bayle writes,

> Sobriety and humility recommend men to God, of whatsoever religion they be, and that the Pagans themselves had a veneration for Christian bishops who testified by their manner of living, that they did not seek any temporal advantage (*DHC*, Marcellinus (Ammianus), B).

But this in turn may be a backhanded compliment, for in praising the bishops of the past it raises the question of whether the bishops of the present likewise forego seeking temporal advantage.

Bayle's *Pensées Diverses* (1682) offers the largest single engagement with the Greco-Roman deities in the philosopher's corpus. The writing is nominally occasioned by Bayle's desire to disprove that the appearance of comets signals divine involvement in the world (Jorink 2008, 54–59). This argument does not remain focussed as Bayle riffs on the nature of faith, the divine, and religious toleration more generally.[26] Infamously, the *Pensées* contains Bayle's assertion that a society of virtuous atheists could exist, thereby separating virtue from right theology.

In a clever argument against comets as signs of the divine, Bayle turns to the history of pagans recognizing comets as supernatural signs. If pagans recognize comets as divine presages, and thus increase in their fervour towards their deities and beliefs then surely the Christian God cannot be operating comets in a supernatural fashion, for there is nothing God despises more than idolatry. The objection to address is that it is better for pagans to believe in false deities, through signs such as comets, than to believe nothing at all. But Bayle will claim here: yes, pagan idolatry is worse than atheism (PD: OD III, 75–77; 83–84).[27]

Far from the often charitable descriptions of Greco-Roman 'pagans' in the *Dictionary*, in the more focussed *Pensées* they often function, at first glance, as foils to good Christian belief.[28] We do see in Bayle a kind of generic dismissal of paganism 'just as' paganism. But this is often couched and cautioned alongside a praising of specific pagans and expression of their virtues. This juxtaposition, of course,

---

[26] Interestingly, in the *Dictionary* article 'Telmessus', on a town full of diviners said to descend from Apollo and possess his gifts, we would expect reference to the *Pensées*. Despite the centrality of interpreting prodigies, Bayle does not do much with this article.

[27] §113;119;129;132.

[28] Although, keep in mind that much of the *Pensées* is devoted to showing how awful 'good Christians' can be.

reflects some consistent themes in Baylean thought: All people can be awful, regardless of religious views. Likewise, all may be good (though Bayle's Calvinist anthropology keeps the pessimistic slant frontloaded!).

Bayle parses the Augustinian solution to the 'problem of paganism' by navigating between the concepts of virtuous action and meritorious action (PD: OD III, 94). Pagans can perform virtuous actions, yes, but because those actions are not directed by a good principle (God) and towards a good end, they are not merited to the pagan as a virtue. It's quite obvious how he then makes the jump to support the virtuous atheist at this point, they are no worse than the pagan insofar as they lack direction by the good principle. Virtue can be performed but without merit. The more intriguing element of this discussion is Bayle's reading back of the pagan religion into his contemporary Christianity. If the Greeks and Romans can be faulted for not being properly directed by a good principle towards a good end, well, so can today's Christian believers! Should Christian actions be deemed meritoriously virtuous even when guided by the same attributes as pagans?[29] The payoff of this challenge is a disruption of the Augustinian framework. We might maintain that Christian actions remain meritorious because of some divine rubber stamp of merit, but can hardly claim that they are always properly ordered when 'pagan' attributes are just as apparent. The reader is pushed to consider whether virtue is as it appears and merit exists as honestly within pagan religions, and apart from religions, as within Christianity.

This question of virtue is central to the majority of articles dealing with persons and religion of Greco-Roman culture. It appears when people behave unexpectedly, displaying virtue their religion does not prescribe. It appears when the gods behave expectedly, for flawed people, but in hardly a divine manner. The relation to the contemporary church and Christian ethic is made explicit at times with Bayle drawing comparisons both positive ('our God is not like *that*') and negative ('Christians today behave similarly'). Though these selected articles are grouped by various thematic focusses, that so many of them intersect and could be arranged differently show the commonalities are deep in the religion discourse of the *Dictionary*. 'Xenophanes' discusses false gods as well as divine sovereignty, 'Scamander' details scandal as well as divine imposters. A consistent group of concerns appear around virtue and religion.

Virtue is inseparable from larger Baylean concerns around the nature of divinity and what is expected of human behaviour, towards each other and towards the divine. The Greco-Roman articles allow Bayle to bring questions towards contemporary religious ethics while maintaining a safe degree of separation. But their comparative function can serve multiple ends depending on the need, or current focus,

---

[29] "mais je soûtiens qu'il n'y a rien là, que l'on ne puisse attribuer au tempérament, à l'éducation, au désir de la gloir, au goût que l'on s'est fait pour une sorte de réputation, à l'estime que l'on peut concevoir pour ce qui paroît honnête et louable, et à plusieurs autres motifs qui sont de la compétence de tous les hommes, soit qu'ils aïent une Religion, soit qu'ils n'en aïent pas" (PD: OD III, 94).

of the author. They contribute to the messy but compelling world of religion in the *Dictionary*, resisting easy answers and dogmatic approaches.

Bibliography

Bayle, Pierre. 1727–1731. *Oeuvres diverses*, 4 vols. The Hague: P. Husson et al.
———. 1734. *The dictionary historical and critical*, 2nd ed. trans. P. Desmaizeaux. London: Knapton et al.
Brockliss, Laurence. 2013. Starting-out, getting-on and becoming famous in the eighteenth-century Republic of Letters. In *Scholars in action*, ed. André Holenstein, Hubert Steinke, and Martin Stuber, 71–100. Leiden, The Netherlands: Brill.
Cole, Spencer. 2014. *Cicero and the rise of deification at Rome*. Cambridge: Cambridge University Press.
Cotton, Parker. 2020. Curious, useful and important: Bayle's 'hermaphrodites' as figures of theological inquiry. In *Exceptional bodies in early modern culture*, ed. Maja Bondestam, 123–140. Amsterdam: Amsterdam University Press.
Gay, Peter. 1966. *The enlightenment*. New York: Norton.
Goodman, Martin. 2008. *Rome and Jerusalem*. London: Penguin.
Jorink, Eric. 2008. Comets in context. Some thoughts on Bayle's Pensées Diverses. In *Pierre Bayle (1647–1706), Le philosophe de Rotterdam: Philosophy, religion and reception*, ed. Wiep van Bunge and Hans Bots, 51–68. Leiden, The Netherlands: Brill.
Jurieu, Pierre. 1686–1689 *Les lettres pastorales aux fidèles qui gémissent sous la captivité de Babylone.*
La Mothe le Vayer, François de. 1642. *De la vertu des Païens*. Paris: Francois Targa.
Marenbon, John. 2017. *Pagans and philosophers*. Princeton: Princeton University Press.
Moréri, Louis. 1674. *Grand dictionnaire historique*, 2 vols. Lyon: Girin and Rivière.
Salmon, John H.M. 1987. Cicero and Tacitus in sixteenth-century France. In *Renaissance and revolt*, 27–53. Cambridge: Cambridge University Press.
Smet, De, and Daniel. 2021. Pythagoras' philosophy of unity as a precursor of Islamic monotheism. Pseudo-Ammonius and related sources. In *Brill's companion to the reception of Pythagoras and Pythagoreanism in the middle ages and the renaissance*, ed. Irene Caiazzo, Constantinos Macris, and Aurélien Robert, 277–295. Brill: Leiden, The Netherlands. p. 288.
Taylor, Helena. 2017. *The lives of Ovid in seventeenth-century French culture*. Oxford: Oxford University Press.
———. 2020. The quarrel of the ancients and moderns. *French Studies* 74 (4): 605–620.
Tinsley, Barbara Sher. 2001. *Pierre Bayle's reformation*. Selinsgrove, PA: Susquehanna University Press.
van der Lugt, Mara. 2016. *Bayle, Jurieu, and the Dictionnaire Historique et critique*. Oxford: Oxford University Press.
Vermeir, Koen. 2012. The dustbin of the Republic of Letters: Pierre Bayle's "Dictionaire" as an encyclopedic palimpsest of errors. *Journal of Early Modern Studies* 1 (1): 109–149.

Chapter 7
Bayle and the Ghosts of Mani and Zoroaster

Jean-Luc Solère

Abstract This chapter re-examines the status and role, in Bayle's reflection, of the dualistic objections to the goodness of a monotheistic God. Bayle's conclusion that these objections are insoluble by reason alone has been taken as an indication that his aim was to undermine an essential tenet of biblical revelation, with the result that his turn to faith to solve the problem of the existence of evil rings hollow. A closer examination of the historical context and logic of these objections, however, reveals that the charge of dualism (especially Manichaeism) was commonly leveled against the Calvinist doctrine of predestination by Catholics and other Protestant denominations throughout the sixteenth and seventeenth centuries. On a purely philosophical level, Bayle could have drawn the dualistic objections he needed from Plutarch, whom he knew well and quotes in the same discussions. If he invokes the ghosts of Mani and Zoroaster in his turn, it is to prove that the accusation of Manichaeism directed against Calvinist orthodoxy can be reversed against the attackers. The opponents' own answer to the problem of evil, which in all its variants rests on the responsibility of the human will, does not solve the problem and exposes them no less than the Calvinists to the charge of making God the author of sin. This strategy is the reason why Bayle places so much emphasis on the question of free will in these discussions. His intention, however, is not to draw out this kind of mutual finger-pointing, but, on the contrary, to prove that these controversies are sterile and endless, and to suppress them as they do more harm than good to Christianity.

Keywords Dualism · Evil · Free-will · Predestination · Religious controversies · Theodicy · Tolerance

J.-L. Solère (✉)
Philosophy Department, Boston College, Chestnut Hill, MA, USA
e-mail: solere@bc.edu

© The Author(s), under exclusive license to Springer Nature Switzerland AG 2024
M. García-Alonso, J. C. Laursen (eds.), *The Importance of Non-Christian Religions in the Philosophy of Pierre Bayle*, International Archives of the History of Ideas Archives internationales d'histoire des idées 251, https://doi.org/10.1007/978-3-031-64865-6_7

In discussions staged by Bayle, representatives of middle-Eastern dualistic religions, namely, Zoroastrians[1] and Manicheans, play an essential role as relentless objectors to the explanations for the presence of evil in the world that Christian philosophies and theologies or, more generally, monotheistic religions traditionally offer. They show that if God is the sole creator of all things, then he is ultimately responsible for all the evils that plague his creation. First summoned by Bayle in his *Dictionary*, they continued to haunt the debates that opposed Bayle to, notably, Jean le Clerc and Isaac Jaquelot, in *Réponse aux Questions d'un Provincial* and *Entretiens de Maxime et de Thémiste*.[2] Readers and critics of Bayle often have suspected that these unlikely antagonists were in fact Bayle's surrogates, and that, hiding behind them, he was expressing his own rejection of the idea that a good God would have allowed evil in the world, that is to say, his rejection of the Christian creed.[3] What arouses this suspicion is that Bayle insists that their objections cannot be repelled by reason alone.[4] They can only be rejected, he says, by unwavering adherence to biblical revelation, which affirms that evil was introduced into creation by man, not by God. One should therefore give up trying to justify God by purely rational means, Bayle concludes, and simply accept what the Holy Writ says.[5] But it seems odd to endorse a view that reason has shown to be objectionable. How could we believe what we *know* to be false? Perhaps there is a path for faith where reason cannot venture and must remain agnostic, but not where reason has firmly barred the way. Because of the apparent implausibility of Bayle's solution, one may think that he is not being straightforward in proposing it, and that he is in fact using Manicheans and Zoroastrians as strawmen to formulate arguments that radically undermine religious beliefs, without, out of caution, explicitly endorsing them.

Thus goes the usual story about Bayle. Here, however, I will attempt to offer an alternative explanation for Bayle's use of these mock proponents of dualism. Placing his choice in the context of the religious controversies of his day sheds light on his motivations for invoking the spirits of Mani and Zoroaster.

---

[1] If not Zoroaster himself, notably in DHC, Manichéens, D, at least "a disciple of Zoroaster" (RQP: OD III, 846—see also "Réponse pour M. Bayle...": OD III, 997–998 and 1002, and EMT: OD IV, 71), or "a Zoroastrian philosopher" (EMT: OD IV, 70–84).

[2] On that debate, see Hickson 2016.

[3] Le Clerc, *Bibliothèque Choisie*, IX, 10: "he attacked the Christian religion from under the mask of Manicheism" (quoted and transl. by Hickson 2016, 58).

[4] DHC, IIe Ecl. (IV, 630): "the Manicheans' objections are insolvable when they are being discussed only within Reason's court of law" (translations of Bayle's texts are mine).

[5] Bayle's response is in fact more nuanced and does not boil down to "blind fideism", as it is sometimes called. See here farther section 7.8, at fn. 84.

7.1 Dramatis Personae

Next to nothing is known about the historical Zaraθuštra (original form of his name, or close to it, as Nietzsche knew, whereas "Zoroaster" derives from its Hellenized and Latinized form). It is not even certain whether he is merely a mythic figure portrayed in the Avesta (the corpus of Zoroastrian sacred texts), or whether he was a real person. If the latter, the proposed dates for his life span vary widely, from sometime in the second millennium BCE, to around the sixth century BCE—in any case within the Iranian cultural sphere, which is about the only certainty we have. Be that as it may, the fact is that he was considered as the founding prophet of a new religion (Zoroastrianism, or Mazdeism as its adherents prefer to call it), which became the established religion of the successive Iranian empires until the Islamic conquest, from the Achaemenids in the sixth century BCE, through the Parthians (247 BCE to 224 CE), to the Sassanids (224 to 651 CE). The core belief of this religion, in its most well-known form, is the cosmic conflict between Ohrmazd (Ahura Mazda), the source of all goodness, and Ahreman, the cause of all evils.[6] Their battlefield is the world, spiritual and material; as a consequence, in it good and evil are inextricably intertwined. As strong as this duality is, however, Goodness will eventually triumph (Bayle omits that part of the Zoroastrian faith, if he was aware of it).[7]

Mani (Manès or Manichaios/Manichaeus in Greek and Latin) was another Iranian prophet, in the third century CE, who created a religion that includes Zoroastrian themes, but also Christian and Gnostic (and perhaps Buddhist) ones, and thus aims at being a grand synthesis. Mani presented himself as "the seal of the prophets", that is, the last prophet who brings to completion the revelation begun by Moses, Jesus, and Zoroaster. Indeed, probably in order to have his creed accepted by the Zoroastrian Sassanian rulers, Mani made sure to present it as the continuation of the old Iranian religion[8] (he did not succeed in convincing them and eventually died in prison). Like Zoroaster, Mani postulates the existence of two conflicting principles, one divine and good, the other demonic and evil. However, his dualism is even more pronounced since, like the Gnostics, he equates matter with evil. In particular, contrary to Zoroastrianism, for Manicheism the creation of humans is the work of demons, intent on imprisoning the World Soul in bodies (an idea that probably comes from Gnosticism). The promised end of the story, however, is the destruction of the material world and the liberation of all the fragments of the World Soul, returning to their place of origin, the world of light.

---

[6] DHC, Arimanius. For a presentation of Zoroastrian dualism, see Skjærvø 2011.
[7] See Duchesne-Guillemin 1998: "Its other salient feature, namely dualism, was never understood in an absolute, rigorous fashion. Good and evil fight an unequal battle in which the former is assured of triumph. God's omnipotence is thus only temporarily limited."
[8] See Sundermann 2009: "The persons of the East Manichean pantheon and pandemonium are figures of a Gnostic system in Zoroastrian garb. (…) Manicheism, one may conclude, is a Gnostic religion with Christian roots and additional Zoroastrian components." (2009). See also Gnoli 1985.

At first blush, Bayle's seizing upon these two ancient dualist religions is intriguing. Manicheism was a confession long since extinct in Western Europe. After successfully spreading in the Roman world, it was suppressed from the end of the fourth century CE on.[9] A last resurgence of it (the so-called Cathar or Albigenses heresy) was crushed in the thirteenth century in Southern France and Italy, after which nobody openly professed this doctrine. As for Zoroastrianism, although it somewhat subsisted (and still subsists), mainly in Iran and India, it lost most of its adherents after the Islamic conquest of Iran, and was never imported as a religion in the West, except perhaps under the derived form of the mystery cult of Mithra. Mithra originally was a Zoroastrian god, but to what extent the cult centered on him was really a branch of Zoroastrianism, or whether it was a later reinvention, is debated. At any rate, Mithraism was also suppressed when the Christians came to power, never to reemerge. It is therefore strange that Bayle summons, as spokespersons for his objections against rationalizing the existence of evil in a monotheistic framework, representatives of these historically and geographically distant religions, which no longer were real challenges or viable alternatives to Christian orthodoxy.

As far as Zoroaster is concerned, however, the inclusion of his fabled figure in Western works was in fact nothing new, but part of a long tradition that began in Antiquity (Bidez and Cumont 1938) and continued long afterward, through Plethon, Ficino, and Patrizi in the Renaissance, to Voltaire, to Mozart's Sarastro in *The Magic Flute*, and at least Melville and Nietzsche, among others (Stausberg 2005).[10] This exotic persona, like some others (such as the Chinese philosophers in Malebranche, Leibniz, or Bayle himself), was sufficiently shadowy, as to its actual teachings, to allow for the projection of Western fantasies, concerns, or interests. As Roger Beck nicely puts it, Zoroaster was to the ancient Greeks little more than a "peg on which to hang home-grown Greek philosophy or other forms of learning and so give it a patina (to change the metaphor) of authority derived from the far away and the long ago. The Greeks considered the best wisdom to be exotic wisdom or — to use the title of Arnaldo Momigliano's masterly study of the phenomenon — 'alien wisdom'" (Beck 2002).[11] This is also true for the Renaissance and modern times. While Zoroaster was vilified by many authors for being the inventor of magic, and, of course, for being a dualist, others were extolling him as a great astronomer/astrologer and a great sage or a truly inspired prophet, on a par with Hermes Trismegistus, like him a beacon of the ancient wisdom that guided the Gentiles in lieu of the biblical Revelation (Masroori and Laursen 2021, 20–21 and 41). Thus, the appearance of Zoroaster in Bayle's writings is only another occurrence of a long-standing practice.

---

[9] See the timeline of the repression of Western Manicheism in Tardieu 2008, 93–95.

[10] On the interest of the West for Persia (as Iran was called at the time) and the available sources of knowledge, particularly in the seventeenth century, see Masroori and Laursen 2021.

[11] Aristotle, in a fragment of the lost *Perì philosophías* (*apud* Diogenes Laertius, 1.8), mentions the doctrine of the "Magi," whose philosophy is older than the Egyptians, and according to which there are two principles, a good demon, Oromasdes, whom he assimilates to Zeus, and a bad demon, Areimanios, assimilated to Hades.

However, Bayle does not fall into any kind of Zoroastermania or Zoroasterphobia. For him, Zoroaster was neither a diabolic magician or an idolater, nor a figure of the eternal wisdom. To his credit, Bayle, as a historian, despite wrongly making him a "king of the Bactrians," does his best to dispel the "colorful tales" (DHC, Zoroastre) that, for better or worse, have accompanied Zoroaster's reputation in the West. In particular, he challenges fanciful claims about the prophet's antiquity (5000 years before the Trojan War, or 6000 years before Plato's death), or his being the same person as Cham, son of Noah, or Nimrod, or Abraham, or Ezechiel, among others (DHC, Zoroastre. DHC, Cham, B). He also denies that he was the creator of the dark arts, and, citing Plato, maintains that, rather than a magician, he simply was a student "of divine nature and religious cult"—that is, in sum, a philosopher-theologian who has sufficient standing to be an opponent in the debates Bayle wants to set up (DHC, Zoroastre; RQP: OD III, 567).

On the other hand, Bayle's appeal to Manicheans as worthwhile sparring partners for Christian orthodoxy is unprecedented, given the revulsion that their very name generally provoked in the Christian world, especially since St Augustine's condemnation and refutation of their doctrine. Manicheism was synonymous with a monstrous theory that denies the unity, goodness, and omnipotence of God. Moreover, while Bayle, as I have just noted, is rather neutral with respect to the person of Zoroaster, he does not in fact distinguish between Zoroastrianism and Manicheism, even though he is aware that the former long predates the latter. In his view, Manicheism simply "renewed (...) one of the most fundamental dogmas" of the Zoroastrians (DHC, Arimanius). Bayle also does not distinguish Manicheism from other Gnostic cults, since in the articles devoted to the Marcionists and the Paulicians, who were Christian Gnostics rather than members of the Manichean church, he has them speak just like the Manicheans.

Of course, this is not mere confusion on Bayle's part: he has a reason for grouping these different religions under the umbrella of Manicheism. For the purpose of the discussions he wants to stage, Bayle is interested in the most radical form of dualism, which is the one that characterizes the Manichean doctrine. Gnosticism can be seen as offering a moderate, non-primordial form of dualism, inasmuch as the evil principle is a creature of the one and only God, but has turned against him (much like Lucifer in traditional Christianity, except that the Gnostic evil principle does much more than tempt human souls: it builds the material world [DHC, Pauliciens, H][12]). Likewise, there is a version of Zoroastrianism (perhaps the most ancient) according to which duality is not originary, evil resulting from a *choice* made by one of the two primeval Twin Spirits, above which is the unique God, Ahura Mazda.[13] This "ethical" dualism (i.e., resulting from a choice) would be of no

---

[12] See Puech 1979, 140–141.

[13] "It can be assumed that the gathic formulation (of Ahura Mazdā and opposed twin spirits) was succeeded by a formulation in which Ahura Mazdā was directly opposed to the evil spirit (...)", which means that "God is degraded to the level of devil's antagonist" (Gnoli 1996). As the author specifies, this simplified form is found much later than the *Gathas*, in the Pahlavi texts. I borrow from Gnoli's article the designations of "ethical" versus "ontological" dualism.

use to Bayle, for one could raise the exact same objections against it that his "Manicheans" raise against Christian orthodoxy (see below): how could a good God have made a creature he would allow to cause evil? DHC, Zoroaster, F. What Bayle needs as a consistent alternative solution to the problem of the origin of evil, to oppose it as a foil to the Christian one, is an extreme, ontological, essentialist dualism, in which two coeternal entities are forever pitted against each other by their very nature, not by anyone's decision.[14]

I will henceforth call this view, which postulates the existence of two independent, eternal, and opposed universal principles or first causes, "diarchism"[15] (and, using Bayle's own term, I will call the opposite view, which recognizes only one principle, "unitarism"[16]). "Diarchism" (two principles, equally uncreated and powerful) is a more precise term than "dualism," which can be applied to conceptions for which the evil principle has somehow split off from the supreme deity. It is the diarchist reading of Zoroastrianism that Bayle defends against Thomas Hyde's *Historia religionis veterum Persarum*, published in 1700, three years after the first edition of the *Dictionary*, in which the Oxford Orientalist, perhaps in reaction to Bayle's use of Zoroaster in the article "Manicheans," describes Zoroastrianism as being unitarist and decries the "dualistic leaners" as a heretical splinter group. Bayle responds in the article "Zoroaster" inserted in the second edition of the *Dictionary*

---

[14] DHC, Pauliciens, H: "We [the Manicheans] are not obliged to seek the cause which makes our bad principle wicked; for when an uncreated thing is such or such, one cannot say why it is so; such is its nature, one necessarily stops there. But as for the qualities of a creature, one must seek their reason, which can only be found in its cause." For the same reason, as noted above Bayle, if he knew about it, had to remain silent about the "happy ending" of the cosmic struggle announced by Zoroastrianism as well as by Manicheism. As his reaction to Le Clerc's "Origenist" solution (salvation for all souls eventually) shows, Bayle was not willing to count on the eventual restoration of the right cosmic order to exonerate the good principle from its shortcomings.

[15] One cannot write "ditheism", since, as Bayle rightly recognizes, Manicheans wouldn't call the evil principle a "God" (DHC, Pauliciens; however, in DHC, Arimanius, Bayle does say that for the Persians, the evil principle was a divinity). Cf. Faustus's protest in Augustine, *Contra Faustum*, book XXI, 1, ed. Zycha p. 568, 9–569, 18: "It is true, we believe in two principles; but one we call God, and the other Hyle, or, to use common popular language, the devil. If you think this means two gods, you may as well think that the health and sickness of which doctors speak are two kinds of health, or that good and evil are two kinds of good, or that wealth and poverty are two kinds of wealth. (...) Do you think that we must call them both gods because we attribute, as is proper, all the power of evil to Hyle, and all the power of good to God? If so, you may as well say that a poison and the antidote must both be called antidotes, because each has a power of its own, and certain effects follow from the action of both. (...) I grant that we, too, sometimes call the hostile nature God; not that we believe it to be God, but that this name is already adopted by the worshippers of this nature, who in their error suppose it to be God. Thus the apostle says: The god of this world has blinded the minds of them that believe not. He calls him God, because he would be so called by his worshippers; adding that he blinds their minds, to show that he is not the true God."

[16] Not to be confused, of course, with the Christian confession called "Unitarianism" because it negates the Trinity.

(1702), remark F, and goes at great length to prove that, on the contrary, Zoroaster was a diarchist.[17]

However, given that there is no doubt whatsoever that Manicheism proposed diarchism in its purest form, and given that the Manichean doctrine was much better known than the Zoroastrian one, due to its entanglement with the religious history of the West and the numerous expositions and refutations of it by the Fathers of the Church, it is only natural that Bayle should gravitate towards Manicheism as the main representative of what he was looking for, and include Zoroastrianism and Gnosticism in it. In fact, it is in the article "Manicheans" that Bayle first introduces "Zoroaster" as an objector and makes him argue against unitarism. And overall, in the *Dictionary* as well as in Bayle's subsequent works, the number of occurrences of the "Manicheans" as proponents of the objections is much greater than that of "Zoroaster" or "Zoroastrians." Likewise, what follows here applies equally to Zoroastrianism and Manicheism.

Let us now turn to the arguments that Bayle draws from diarchism.

7.2 Why Diarchism?

The core of the *Dictionary*'s discussion of evil is a group of three articles: "Manicheans", "Marcionites," and "Paulicians". But as Mara Van der Lugt has noted, the *Dictionary*'s system of cross-references expands the discussion widely through a constellation of interrelated articles (Van der Lugt 2016). This web of cross-references reveals Bayle's intense concern with the problem of evil (Jossua 1977).

Only human beings are a source of difficulties for unitarism, for the imperfections of natural processes among purely corporeal things are easily explained by Malebranche's theory of the simplicity and immutability of God's action and of occasional causes (DHC, Manichéens, D). The diarchist path starts with the observation that human lives are a mixed bag of good and bad things. The bad, in human lives, is twofold: "Man is wicked and unhappy" (DHC, Manichéens, D), that is to say, on the one hand, human beings are evil and commit evil acts, and on the other hand, they suffer (not only physically, but also psychologically). In Bayle's dark outlook, most of human history consists of the various crimes committed by human beings and the misfortunes that befall them (the consequences of these crimes, but also natural disasters, from famines to pandemics to earthquakes, for example).[18]

---

[17] DHC, Zoroastre: "the presumption is great that he actually taught that there are two coeternal causes, one of good things, the other of evil." Of course, seeing Zoroastrianism as unitarian made it easier to find in it a token of the true, natural, and rational religion, as some of the Cambridge Platonists did (see Masroori and Laursen 2021, 41). By presenting Zoroaster as a diarchist, Bayle makes him unassimilable to any rational religion, or Christianity for that matter.

[18] Ibid.: "Man is wicked and unhappy: everyone knows it by what he feels in himself, and by the dealings he is obliged to have with his fellow human beings. You only need to live five or six years

But there are also moments of happiness and virtuous actions in human life.[19] It is this duality of good and evil, happiness and misery that, according to Bayle, led Zoroaster and Mani to postulate two eternal principles at odds with each other.[20]

The introduction of a competing evil principle singularized their religious thought and made them the target of relentless attacks from traditional, unitarian theologies and philosophies. Bayle, however, believes that they will always have the upper hand in a purely philosophical discussion of the origins of evil. According to him, unitarists are at a loss when they try to explain why evil exists. All their attempts can be met with objections they are unable to repel.

7.3 A Failed Solution: Man's Free Choice

In order to drive his point home, Bayle stages a debate between Zoroaster and Melissus, the fifth century BCE Presocratic philosopher. The choice of the latter as Zoroaster's interlocutor is a bit odd, given that Bayle has him say things that the real Melissus could not have said (more on this in a moment). But, first, at this stage of the debate Bayle needed a pagan philosopher rather than a Christian one who could appeal to Revelation, whereas here Bayle wants to keep the debate purely philosophical. Second, Melissus is a representative of the extreme opposite of cosmic dualism, namely, a representative of the most radical form of monism, since he was a student of Parmenides and held the Eleatic view that Being is one, indivisible, non-generated, and unchanging.

---

to be perfectly convinced of these two points (...). Travels provide permanent lessons on this subject: Everywhere you go, you see testimonies of man's misfortune and wickedness; everywhere, prisons and hospices; everywhere, gallows and beggars. (...) Strictly speaking, History is nothing but a compendium of the crimes and misfortunes of the human race." See also DHC, Xénophanes, E.

[19] Bayle, though, ever the pessimist, does not think that the amount of happiness is even close to the amount of unhappiness in human lives. See DHC, Panormita, H; Périclès, K; Raphelengius, B; Xénophanes, F and K; and Van der Lugt 2016, 37–39.

[20] DHC, Manichéens, D: "(…) but let us note that these two evils, one moral and the other physical, do not occupy the whole of History, nor the whole experience of individuals. We find everywhere both moral and physical good: some examples of virtue, some instances of happiness. This is what creates the difficulty. For if there were only the wicked and the unfortunate, we would not need to resort to the hypothesis of the two Principles. It is the mingling of happiness and virtue with misery and vice that requires this hypothesis; this is where the strength of Zoroaster's sect lies. » Also, DHC, Pauliciens, E: "Therefore, you cannot explain our experiences except by the hypothesis of the two principles. If we feel pleasure, it's the good principle that gives it to us; but if we don't feel it pure, and are soon tired of it, it's because the bad principle interferes with the good one. The good principle reciprocates: it makes it so that pain becomes less felt through habituation and that we always have some resource left in the face of the greatest evils. This, and the good use one often makes of adversity, and the bad use one often makes of happiness, are phenomena that are admirably explained by the Manichaean hypothesis. They induce us to suppose that the two principles have entered into a transaction that reciprocally limits their operations. The good cannot do us all the good it wishes to. In order to do us as much good as it does, it had to consent to its adversary causing the same amount of harm."

In this debate, Zoroaster challenges Melissus to provide an explanation for the existence of evil that is consistent with his monistic framework:

> If man is the creation of a single Principle, sovereignly good, sovereignly holy, sovereignly powerful, can he be exposed to disease, cold, heat, hunger, thirst, pain, sorrow? Can he have so many evil inclinations? Can he commit so many crimes? Can sovereign holiness produce a criminal creature? Can sovereign goodness produce an unhappy creature? (DHC, Manichéens, D)

Melissus's answer is that (it's here that Bayle's Melissus diverges from what the real Melissus could have thought) God created man in a state of happiness and innocence, but man strayed away from the right path and became morally evil. As a result, God punished him by introducing physical evil, which is compatible with God's goodness because it is a just punishment and an expression of another of God's essential attributes, justice. Thus, man's suffering and inclination to evil are the consequences of his own fault; God is in no way responsible for them (DHC, Manichéens, D).

This account obviously follows the outlines of the story of Adam's and Eve's fall in the Bible. The episode, as usually interpreted by Christian theology, is supposed to explain the presence of evil in the world in the same way, that is, by placing the entire responsibility for it on human beings. Accordingly, the Christian response focuses on the freedom of the human will. Since Adam and Eve were given free will and had the possibility of not succumbing to temptation, they alone are the cause of the Original Sin, and from this first evil act flow all other sins as well as physical evils, which are a retribution (death, sickness, suffering, etc.). Therefore, when Zoroaster refutes Melissus's account, he already strikes at a core tenet of the Christian narrative regarding the introduction of evil into God's creation, even though he is supposed to be discussing with a pagan philosopher. He thus anticipates discussions with Christian thinkers, which Bayle will develop in other articles, such as "Marcionites" and "Paulicians," that relentlessly return to the idea that our free choice does not absolve God of his responsibility regarding the existence of evil.

The gist of Zoroaster's response, as well as that of the Manicheans enrolled by Bayle in subsequent discussions, is that when God gave man the possibility of choosing evil, he must also have foreseen that man would make this original evil choice.[21] If so, it is absolutely evident to our "natural light", that is, our reason, that

---

[21] Bayle gets perfectly right what real diarchists did answer. See how the ninth century Zoroastrian author Mardan-Farrukh (about whom Bayle could not have had any knowledge) argues in his *Shkand-gumanig Vizar* (*Decisive Solution for Doubts*) as summarized by (Skjærvø 2011: 74–75): "Going on the offensive, Mardânfarrokh son of Ohrmazddâd (...) points out the impossibility that an omniscient, omnipotent, good, and merciful divine being should produce a being so contrary to his creator. And if the being he created was not made this way, but became evil afterward, it would seem that the evil creature was more powerful than its creator. Similarly, against the Jews, why did the creator create Adam and Eve if he knew beforehand that they would turn away from his will and force him to punish them? If he did not know beforehand, then, surely, he must be ignorant and badly informed. Against the Christians, as for the free will of men, since it was made for them by the creator himself, but was then used to commit sins, then god himself is logically the originator

God either should have prevented that choice, or should not have given the possibility of choice in the first place. It is a fundamental ethical principle that one is responsible for the harm that one has caused, even indirectly, and for the evil one could have prevented from occurring. These two conditions are met in the case in point: it is God who gave man the possibility of choosing, and he could easily have prevented man from making the wrong choice. The second clause results from the attributes of omniscience and omnipotence traditionally ascribed to God. Because of his omniscience, the future is not hidden from him and he knew in advance that Adam and Eve would abuse their freedom.[22] Because of his almightiness, he was undoubtedly able to stop them; nothing could have made that impossible. Now, common sense tells us that if we are certain that the potential recipients of a gift would use it to harm themselves or others, we should not provide that gift. Otherwise, we would be morally responsible for the consequences, just as if we had given a loaded gun to a child. The reality of our love for the recipient of the gift could also be called into question.[23]

Bayle offers several quaint similes to dismiss all the reasons why God might have allowed sin to happen, that is, might have allowed Adam and Eve to do with their freedom what he knew they would do. Here are some examples:

---

of those sins, since it was he who gave them the free will in the first place." See also Jossua 1977, 92–93.

[22] DHC, Pauliciens, F: "If you say with those who have come closest to the method that would exonerate Providence, that God did not foresee the fall of Adam, you gain but little; for at the very least he knew with great certainty that the first man would run the risk of losing his innocence and of introducing into the world all the evils pertaining to affliction and guilt, as a result of his rebellion. Neither God's goodness, nor his holiness, nor his wisdom, may have allowed him to risk these events (...). [The Socinians] cannot say that he knew the sin of the first man only as a possible event. He knew all the steps of temptation, and he must have known a moment before Eve succumbed that she was going to lose herself. He must, I say, have known it with such a certainty as renders one inexcusable if one does not remedy the evil (...) the Socinian system, depriving God of foreknowledge, reduces him to servitude and to a form of government which is pitiful, and does not remove the great difficulty which should have been removed, and which forces these heretics to deny the foreknowledge of contingent events."

[23] DHC, Pauliciens, E: "Those who say that God allowed sin, because he could not have prevented it without undermining the free will he had given man, which was the most beautiful gift he could have given him, are exposing themselves a great deal. The reason they give is beautiful, it has a dazzling *I know not what*, it is grand; but in the end, it can be countered by reasons that are more within the reach of all men, and more founded on common sense and the ideas we have of order. Without having read Seneca's beautiful treatise on favors, we know by natural light that it is the essence of a benefactor not to offer favors which he knows would be abused in such a way that they would only serve to ruin the person to whom he would give them. Otherwise, there is no enemy so inveterate that he would not shower his adversary with favors." See also DHC, Pauliciens, M: "The same goodness that leads us to give something that we judge capable of making the people who will enjoy it happy, leads us to take it back as soon as we observe that it makes them unhappy; and if we have the necessary time and strength, we don't wait to take away this present until it has already been the cause of unhappiness, we take it back before it has done any harm. This is where we are led by the ideas of order and by the notions by which we can judge the essence, and characters of goodness, in whichever subject it is found, creator or creature, father, master, king, etc."

(a) Would God have allowed sin to happen in order to show his goodness by forgiving the sinners? Everyone knows that it shows greater goodness to prevent a man from falling into a well than to let him fall so that he can be rescued (DHC, Pauliciens, E).
(b) Would God have let sin happen in order to make redemption possible and thus reveal his glory? He would be like a surgeon who allows his children's legs to be broken so that he can demonstrate his skill in putting them back together (DHC, Pauliciens, E).
(c) Would God have allowed evil because he did not want to interfere with the freedom of Adam and Eve? He would then be like a mother who, not wanting to be authoritarian, lets her children go to a party where she is sure they will be involved in drug dealing, and simply urges them not to do anything foolish (I am modernizing Bayle's example: DHC, Pauliciens, E).

Even if a greater good resulted from the existence of evil, the price was too high. God, who was free to create or not to create, should have refrained from creating at all if there was no way to avoid the infinity of evils that he knew would occur. Nothing compelled him to act. Therefore, when he decided to create the humans, the existence of evil was in fact deliberately chosen by him. As a consequence, he is as much as humans the author of sin.

In short, it is mere sophistry to say that God made the sin of Adam and Eve possible and did not prevent it, but that he is not responsible for it. To put it casually, even if it is true that sinning was due to the first humans' choice, the buck stops with God. Even if he did not decide by antecedent will (that is, prior to the humans' choice, as in the most extreme Calvinist view) that they would sin, at the very least he had to consent to their sin by consequent will (that is, after foreseeing what their choice would be, as in the Catholic and moderate Protestant views), which means that he had to decide that the history of his creation would unfold in that manner. Therefore, one way or another, he wanted them to sin.[24] In short:

> (…) the manner in which evil was introduced under the government of a sovereign being infinitely good, infinitely holy, infinitely powerful, is not only inexplicable, but even incomprehensible; and everything that can be opposed to the reasons why this being would have

---

[24] DHC, Pauliciens, F: "(…) it is impossible to understand how a mere permission can make pure possibilities become contingent events, or how it can put the Divinity in a position to be certain that the creature will sin. Mere permission is no ground for divine foreknowledge. This is why most theologians assume that God has decreed that the creature will sin. According to them, this is the basis of foreknowledge. Others want the decree to imply that the creature will be placed in the circumstances in which God foresaw that it would sin. So some want God to have foreseen sin because of his decree, and others want him to have made the decree because he had foreseen sin. Whichever way one explains it, it clearly follows that God wanted man to sin, and that he preferred this to the perpetuation of man's state of innocence, which it was so easy for him to procure and order. Accord this, if you can, with the goodness he must have for his creature, and with the infinite love he must have for holiness."

permitted evil, is more in accordance with natural light and the ideas of order, than these reasons.[25]

As a result, the God who allowed sin to happen does not seem to be a good, just, and moral God. What our reason tells us a good and just being should do does not match what the Bible tells us about God's actions.[26] On the contrary, God seems to be evil. His decision betrays a monstrous cruelty rather than any goodness, since the first humans' sin has caused and will cause the infinite suffering of an infinite number of people, most of whom will be damned.[27]

Consequently, diarchists can legitimately claim that unitarism implies that its only God is an evil God. If, on the other hand, one supposes that the true God is wholly good, then one must admit that there is an independent cause from which all evils derive. To admit this, as Bayle says, is only to abandon a "less reasonable Manicheism," which posits good and evil in one and the same God, "which is monstrous and impossible," for a "more reasonable Manicheism," which keeps the principles of good and evil distinct from each other.[28] The fall of Adam and Eve, in this view, is the result of the independent action of the evil principle, leaving the good principle uninvolved in and uncompromised by their sin.

---

[25] DHC, Pauliciens, E. The difficulty is specific to the Abrahamic, monotheistic religions, which posit God as an absolute endowed with these attributes of almightiness, infinite goodness, wisdom, and prescience, whereas, as Bayle notes in remark G of the same entry, ancient pagan religions had an easier way of explaining evil without being diarchist, since their gods had limited capacities, were not models of virtue but had passions that involved them in human disputes, etc.

[26] DHC, IIe Ecl. (IV, 635): "The objections it [reason] forms against the mysteries of the Trinity, and of the Incarnation, are usually only felt by those who have some knowledge of logic, and of metaphysics; and as they belong to speculative sciences, they strike less the common man; but its arguments against Adam's sin, and against original sin, and against the eternal damnation of an infinite number of people who could not be saved without an effective Grace that God gives only to his chosen ones, are founded on principles of morality that everyone knows, and which continually serve as a rule for both the learned and the ignorant, to judge whether an action is unjust, or not. These principles are utterly obvious, and act on the mind and heart in such a way that all our faculties are upset when we have to impute to God a conduct that does not conform to this rule." See also DHC, Origène, E, and Simonide, F, and RQP: OD III, 809.

[27] DHC, Pauliciens, F: "It is clear to any reasoning man that God is a sovereignly perfect Being, and that of all the perfections there are none more essentially suited to Him than goodness, holiness and justice. As soon as you take these perfections away from him, and give him those of a lawgiver who forbids man to commit crime, and who nonetheless pushes man into crime, and then punishes him for it eternally, you make of him a nature in whom no confidence can be placed, a deceitful, malignant, unjust, cruel nature (...)". See also DHC, Prudence, F.

[28] DHC, Pauliciens, F: "(...) The one principle that you admit has, according to you, willed from all eternity that man sin, and that the first sin be contagious, that it produce endlessly and ceaselessly every conceivable crime on the face of the Earth; after which he prepared for mankind in this life every conceivable misfortune: pestilence, war, famine, pain, sorrow; and after this life a Hell where almost all men will be eternally tormented, in a way that makes one's hair stand on end when one reads the descriptions of it. If such a principle is, moreover, perfectly good, and if it loves holiness infinitely, must we not recognize that the same God is both perfectly good and perfectly bad, and that he loves vice no less than virtue? But isn't it more reasonable to divide these opposite qualities, and give all the good to one principle, and all the evil to the other?".

7 Bayle and the Ghosts of Mani and Zoroaster

Of course, one might want to respond to Bayle's arguments by saying that once God has endowed human beings with free will, he cannot influence their will without contradicting himself, that is, without undoing what he has done. In granting freedom of choice, God has bound himself not to intervene. In fact, this is what Christian theologies have maintained since the beginning, that is, since the Fathers of the Church had to refute the Gnostics and the Manichaeans.[29]

Examining this response, Bayle laments the poor debating skills of the diarchists (Gnostics like Marcion and the Manicheans themselves) who were unable to provide an answer in this discussion with the Fathers.[30] As a result, he decides to argue on their behalf and show how easily this explanation of why God did not intervene can be refuted. However, Bayle's motivation is not simply to remedy the shortcomings of the ancient diarchists; as we'll see, he has other reasons for bringing the question of free will to the fore, and as we'll also see, these reasons explain why he saw fit to resurrect the diarchist view.

First, Bayle refers to the Christian belief that in the afterlife, the blessed, in what is called the beatific vision or the face-to-face contemplation of God, are in a state of perfect conformity with God's will, a state that is definitive and from which they cannot depart—which is to say that they cannot sin anymore. However, the immutable adherence of their will to the true good does not deprive them of their freedom. At least, no theologian would say that it does, since the beatific vision is a state of perfection, in which nothing that belongs to human nature is lost, but in which, on the contrary, all human qualities are brought to their maximum. Therefore, without violating their freedom, God could have made it so that Adam and Eve would not be tempted.[31] Bayle cunningly adds:

---

[29] DHC, Pauliciens, main text: "The ancient Fathers were not unaware that the question of the origin of evil is not an easy one. They were unable to resolve it by means of the Platonic hypothesis, which was basically a branch of Manichaeism, since it admitted two principles, and they were obliged to resort to the privileges of human freedom. But the more we reflect on this way of untying the difficulty, the more we realize that the natural lights of Philosophy provide the means to further tighten and entangle this Gordian knot."

[30] DHC, Marcionites, F: "The Marcionites are pitiful when they argue about this. Generally speaking, if we assess their strength by the objections they raise in Origen's *Dialogue*, we will have a bad opinion of them. One does not find that they pressed home the difficulties about the origin of evil, for it seems that as soon as they were told that evil had come from the misuse of man's free will, they didn't know what to reply, or that if they based a counter-argument on the [divine] foreknowledge of this bad use, they were satisfied with whichever answer, however weak it might have been."

[31] DHC, Marcionites, F: "Origen having replied that an intelligent creature who wouldn't have possessed free-will, would have been immutable and immortal just like God, he shuts the Marcionite down, for the latter replies nothing. It was, however, easy to refute this answer. All one had to do was ask Origen whether the blessed in Paradise are equal to God with regard to the attributes of immutability and immortality. He would doubtless have replied no. Consequently, he would have been told, a creature does not become God as soon as bound to do good and deprived of what you call free-will. (...) St. Basil made another reply, which has the same flaw. God, he says, did not want us to love him by force, and we ourselves do not believe that our servants are affectionate in their service to us while we hold them in chains, but only when they obey willingly. (...) To convince St. Basil that his thinking is very false, we need only remind him of the situation in Paradise."

> If today there were Marcionites as strong in dispute as either the Jesuits are against the Jansenists, or the Jansenists against the Jesuits, they would begin where their ancestors were ending. They would first attack Origen's last stronghold, namely free will (....).[32]

Second, Bayle argues that theologians recognize that God's grace is compatible with human free will, since it directs free will towards the good without infringing on it. Thus, God could have preserved Adam's and Eve's freedom while preventing them from sinning.

> Does the Grace of God reduce the faithful to the condition of a slave who obeys only by force? Does it prevent them from loving God voluntarily, and from obeying Him with a free and sincere will? If this question had been put to Saint Basil and to the other Fathers who refuted the Marcionites, would they not have been obliged to answer negatively? But what is the natural and immediate consequence of such an answer? Is it not to say that, without offending the creature's freedom, God can infallibly turn it towards the good? Therefore, sin did not come about because the Creator could not prevent it without ruining the creature's free will. One must look for another cause. It is hard to fathom why the Fathers of the Church did not see the weakness of what they were answering, and why their opponents did not point it out to them (DHC, Marcionites, G).[33]

God is loved there, God is served perfectly well; and yet the blessed do not enjoy free will; they no longer have the disastrous privilege of being able to sin. Should we compare them to those slaves who obey only by force? What was St. Basil thinking?"

[32] Ibid. See also Bayle's answer to Lactantius, who had argued that God had to permit evil, otherwise we would not have any knowledge of virtue, wisdom, etc.: "Is there anything more monstrous than this doctrine? Does it not overturn everything theologians tell us about happiness in Paradise, and the state of innocence? They tell us that Adam and Eve in this blissful state felt, without any mixture of discomfort, all the sweetness presented to them by the Garden of Eden, the delightful and charming dwelling place where God had placed them. They add that if they had not sinned, they and all their descendants would have enjoyed this happiness without being subject to disease or sorrow, and without the elements or animals ever being against them. It was their sin that exposed them to cold and heat, hunger and thirst, pain and sorrow, and the harm that some beasts do to us. So, far from virtue and wisdom being impossible for man without physical evil, as Lactantius asserts, it must be argued on the contrary that man was subject to this evil only because he had renounced virtue and wisdom. If Lactantius' doctrine were correct, we would necessarily have to suppose that the good angels are subject to a thousand inconveniences, and that the souls of the blessed pass alternately from joy to sorrow: so that in the abode of glory, and in the bosom of the beatific vision, we would not be safe from adversity. Nothing is more contrary than this to the unanimous sentiment of theologians and to right reason" (DHC, Pauliciens, D).

[33] See also DHC, Pauliciens, M: "With all this, were there not enough means to prevent the fall of man? It was not about opposing a bodily movement: that's a grievous opposition; it was only about an act of will. Now, all philosophers cry out that the will cannot be forced, *voluntas non potest cogi*, and it is a contradiction to say that a volition is forced; for every act of the will is essentially voluntary. But it is infinitely easier for God to imprint in men's souls such acts of will as he sees fit, than it is for us to fold a napkin; therefore, etc. Here's another, even more conclusive observation. All theologians agree that God can infallibly produce a good act of will in the human soul without depriving it of the functions of freedom. A prevenient delectation, the suggestion of an idea that weakens the impression of the tempting object, a thousand other preliminary means of acting on the mind and on the sensitive soul, ensure that the reasonable soul makes good use of its freedom, and turns towards the right path without being invincibly pushed there. Calvin would not deny this with regard to Adam's soul during the time of innocence, and all the theologians of the Roman Church, without excepting the Jansenists, admit it with regard to a sinful man. They

Beyond the Church Fathers, everyone, in modern times, agrees that human free will is not destroyed when God intervenes in the decision it makes.

On the Protestant side:

> Nowadays, the most rigorous Christian sects acknowledge that God's decrees did not impose the necessity of sinning on the first man, and that the most effective Grace does not take away the sinner's freedom. They therefore admit that decrees to keep mankind constantly and invariably in the status of innocence, however absolute they might have been, would have allowed all men to fulfill all their duties quite freely (DHC, Marcionites, G).

And on the Catholic side:

> The Thomists maintain that physical predetermination perfects the freedom of our soul, far from taking it away, or injuring it; nevertheless they teach that this predetermination is of such a nature, that when it is given to produce an act of love, it is not possible *in sensu composito*, for the soul to produce an act of hatred. (...) they provide the means to completely overturn Saint Basil's solution to the objections of the Manicheans; and as for the Molinists, they could not use such a solution; for they do not reject the graces of God which infallibly assure a man of his predestination, nor do they deny that if God willed, he could not make a man always act freely, and always avoid sin in the most perilous temptations (DHC, Marcionites, G).

Thus, with Bayle's help, the diarchists are able to defeat any response to their objections that appeals to human free will. It seems that they are in an inexpugnable position and that only their theory is able to account for the existence of evil without holding God responsible for it.

7.4 The Flaws of Diarchism

Despite the strength of the diarchist position, Bayle's choice of Zoroastrians and Manicheans as antagonists remains curious. As we have just seen, he belittles them because he finds them to have little to no philosophical aptitude. Not only did they add inept details and ludicrous fables to the fundamental insight of their system (not to mention questionable practices),[34] but they were unable to put to use the full potential of their dualistic insight.[35] Why choose as spokespersons figures who are not up to the task of arguing with Christian theologians?

---

acknowledge that his actions can be meritorious, even though he only acts with a grace that is either effective by itself, or sufficient to such a degree that it is infallibly followed by its effect. They must therefore acknowledge that an assistance provided by God to Adam so timely, or so conditioned that infallibly it would have prevented him from falling, would have accorded very well with the use of free will, and would have made no constraint felt, nor anything unpleasant, and would have preserved the opportunity to acquire merits."

[34] DHC, Manichéens, main text and DHC, Manichéens, B; DHC, Zoroastre, E.

[35] RQP: OD III, 1050: "the system of the Manicheans is so confused, so variegated, so disjointed, and so full of absurdities which do not follow from one another, that it is clear that they had no rightness of mind, and that they were very bad logicians." See also DHC, Marcionites, F (III, 318b): "If a man of as much wit as Mr. Descartes had been in charge of this matter, the system of

Moreover, Bayle is clear that the diarchist conception is inconsistent, fragile, open to devastating objections, and easily refuted on points other than the origins of evil. Contrary to Michael Stausberg's assessment (Stausberg 2000, 586), Bayle does not think that the Manichean view is superior to the unitarist view on the whole. It is true that it explains phenomena better by its reasoning *a posteriori*, that is, from the existence of evil to its simplest explanation. But it contains propositions which are contrary to clear and evident ideas, and it does not stand up to *a priori* arguments. The hypothesis of two principles is metaphysically "absurd and contradictory."[36] God, the first cause, is a *per se* necessary being, which must therefore be infinite and, consequently, cannot be limited by a competing first cause.[37] In the discussion imagined by Bayle, Zoroaster himself admits that Melissus's view is more in conformity with the "ideas of order" and is superior in its simplicity (DHC, Manichéens, D). As Bayle explains in the *Second Clarification* appended to the *Dictionary*:

> I said [in the "Manicheans" article] that the quality of a system consists in the fact that it contains nothing repugnant to obvious ideas, and in the fact that it gives an account of Phenomena. I added that the Manichean system has at most only the advantage of explaining several phenomena which strangely embarrass the partisans of the unity of Principle; but that besides this point, it is based on a supposition that is repugnant to our clearest ideas, whereas the other system is based on these notions. By this remark alone, I give superiority to the Unitarians, and I take it away from the Dualists; for all those who are expert in reasoning agree that a system is much more imperfect when it lacks the first of the two qualities I mentioned above, than when it lacks the second. If it is built on an absurd, awkward or implausible supposition, this cannot be compensated by a satisfactory explanation of the phenomena; but if it does not explain them all satisfactorily, this can be remedied by its clearness, its plausibility, and by its conformity to the laws and ideas of order; and those who have embraced it because of this perfection have not let themselves be put off by the pretext that it cannot account for all experiences. They impute this defect to the smallness of their lights. (...) (DHC, IIe Ecl (IV, 638).

Worse, even the diarchist explanation of the origin of evil is exposed to fatal objections, as Bayle again notes again in the *Second Clarification*.

First, it is exposed to the counter-charge that its allegedly good God is actually evil. In the Manichean cosmology, the Good Principle is reduced to making a

---

the two Principles could not have been refuted so easily as it was by the Fathers, who had to contend only with a Cerdon, a Marcion, an Apelles, a Manes, people who were not able to make good use of their assets, either because they admitted the Gospel, or because they were not sufficiently enlightened to avoid the explanations most prone to great inconveniences."

[36] DHC, Pauliciens, E: "Who will not wonder at and deplore the destiny of our reason? Here are the Manicheans who, with a completely absurd and contradictory hypothesis, explain experiences a hundred times better than the Orthodox do with the supposition so just, so necessary, so uniquely true of a first principle infinitely good and omnipotent." See also "Mémoire": OD IV, 179.

[37] DHC, Manichéens, D: "They would soon have been put on the run by a priori reasons (...) The surest and clearest ideas of order teach us that a being that exists by itself, that is necessary, that is eternal, must be unique, infinite, all-powerful, and endowed with all kinds of perfections. Thus, in consulting these ideas, we find nothing more absurd than the hypothesis of two eternal and independent principles (...) These are what I call a priori reasons. They necessarily lead us to reject this hypothesis, and to admit only one principle of all things."

compromise with the Evil Principle in order to end the chaos that their struggle engenders. In a move similar to the compact that, according to Hobbes, ends the state of nature, the two Principles agree to each limit their claims and to mutually recognize the rights of the other on what they have renounced. In this way, they each have a say in the government of the world.[38] But in doing so, the Good Principle commits a grave injustice to the detriment of innocent souls, as the sixth century Neoplatonist Simplicius of Cilicia pointed out:

> The fear he had of an irruption by his enemy, they said, obliged him to give up some of his souls to save the rest. These souls were portions and members of his substance, and had committed no sin. Simplicius concludes from this that it was unjust to treat them in this way, especially given that they were to be tormented, and that, in case they contracted any impurity, they were to remain eternally in the power of evil. (...) The author concludes that by refusing to acknowledge that God is the Author of evil, we have made him evil in every way.[39]

Second, another argument borrowed from Simplicius proves that dualism is self-defeating. If there is a powerful Evil Principle that even the Good Principle cannot overcome, then humans cannot resist it. And if this Evil Principle invincibly pushes them to commit evil actions, they are not guilty of these evil-doings and there is no sin. But if there is no sin, there is no need to suppose a principle of evil. Thus, the dualist hypothesis leads to a contradiction.[40]

Finally, Bayle finds in Arnobius another objection that points to a contradiction in dualistic thinking. Manichaeism and Zoroastrianism presented themselves as religions. But according to their own ideas, the Good Principle can only do good,

---

[38] DHC, Manichéens, D: "Zoroaster would go back to the age of Chaos. It is a state that, with regard to his two Principles, is very similar to that which Thomas Hobbes calls the state of nature, and which he supposes to have preceded the establishment of societies. In this state of nature, man was wolf to man, everything belonged to the first occupant: no one was master of anything unless he was the strongest. To get out of this abyss, everyone agreed to relinquish their rights over everything, so that the ownership of something could be ceded to them. The two Principles, weary of chaos, where each was disrupting and upsetting what the other wanted to do, came to an agreement: each ceded something; each had a share in the production of man, and in the laws of the union of the soul. (...) By means of this agreement, the chaos was untangled; chaos, that is, a passive principle that was the battleground of the two active principles. (...) This is what Zoroaster could allege, boasting that he did not ascribe to the good Principle the production, on its own accord, of a work that was to be so wicked and wretched; but only after he had experienced that he could not do better, nor oppose in a better way the horrible designs of the bad Principle."

[39] DHC, IIe Ecl. (IV, 639). Though, in DHC, Pauliciens, I, Bayle seems to respond to this objection that God does this to avoid a greater evil.

[40] "He says that it [the hypothesis of the Two Principles] entirely overturns the freedom of our souls, and that it requires them to sin, and consequently that it implies contradiction; for since the Principle of evil is eternal and imperishable, and so powerful that even God cannot overcome it, it follows that the human soul cannot resist the impulse with which it is pushed towards sin. But if it is invincibly impelled, it does not commit homicide, adultery, etc., through its own fault, but by a force majeure that comes from outside; and in this case it is neither criminal nor punishable. There is therefore no longer any sin, and so this Hypothesis destroys and exterminates itself, since if there is a Principle of evil, there is no longer any evil in the world; but if there is no evil in the world, it is clear that there is no Principle of evil. (...)" (DHC, IIe Ecl. (IV, 639)).

and the Evil Principle can only do evil. Therefore, there is no point in praying to either of them: the Evil Principle will never stop harming us, and the Good Principle will always provide help, albeit limited, no matter what the believers do. Consequently, these principles cannot be the object of any cult, and the diarchist doctrine is "the eraser of all religions."[41]

It is strange, then, that Bayle should choose to voice his own objections to monotheistic theodicies through systems of thought that are so flawed as to offer no viable alternative.[42]

Our perplexity increases even further when we realize that Bayle could have found in the West more suitable figures to lead his onslaught, as I will now demonstrate.

7.5 Why Not Plutarch?

Bayle notes that there were enough dualistic ideas in ancient Greek philosophy for the Western Gnostics not having needed Manicheism as a source of their worldview. Before the Manichean religion was introduced in the Roman world, they drew their inspiration from some Western pagan philosophers (DHC, Manichéens, C). The same might have been true of Bayle himself: he could very well have enlisted the Platonist philosopher Plutarch of Chaeronea (c. 45–120 CE) to counter the too facile answers given to the problem of evil in a unitarist framework. In the article "Manicheans," Bayle identifies Plutarch not as a mere historian of dualism but as a committed diarchist himself, and explicitly relates his view to that of Zoroastrians (the "Persian philosophers").[43] Likewise, in the article "Paulicians," Bayle mentions

---

[41] DHC, Pauliciens, G: "Although this passage by Arnobius favors the Manicheans, it contains a remark that embarrasses them and overturns their whole cult; for the reason why they admitted an evil principle, was that they did not believe that the good principle could do evil: they therefore believed that the other could not do good; thus all their divine service was useless, the beneficent God would never have punished their irreligion, and they could never have made the maleficent God propitious to them". Bayle repeats the argument in DHC, IIe Ecl., (IV, 639): "(…) the Dogma of the Manicheans is the sponge of all religions, since by reasoning consequently they can expect nothing from their prayers, nor fear anything from their impiety. They must be convinced that whatever they do, the good God will always be favorable to them, and the bad God will always be contrary to them. They are gods such that one of them can only do good, and the other can only cause harm."

[42] It is no less strange that Michael Stausberg should affirm that "Bayle does not refute the Manichaic doctrine at all" (Stausberg 2000, 585). This refutation also shows it is quite unthinkable that Bayle himself might have been a Manichean (the last of the Manicheans), contrary to what, surprisingly, as serious an interpreter as Elisabeth Labrousse seems to suggest as a possibility (Labrousse 1963, 270–271). There is no need to suppose that he endorsed the Manicheans' views to explain why he has recourse to them, as we'll see at the end of this chapter. See also Jossua 1977, 148–153.

[43] DHC, Manichéens, C: "Plutarch is going to inform us about the antiquity and universality of this system, not as a simple historian, but as a faithful sectator. It is impossible, he says, for there to be

him, as well as Plato, along with Zoroaster, Marcionists, and Manicheans, as advocates of diarchism.[44] Admittedly, in order to establish the antiquity and universality of the diarchist view, Plutarch himself brings in Zoroaster as an ancient proponent of the two principles.[45] But he also mentions Heraclitus, Empedocles, Pythagoras, Anaxagoras, Aristotle, and Plato.[46] For Plutarch, the Western tradition, as he understands it, carries no less dualism than the Middle Eastern one. As he says in the passage of *On Isis and Osiris* that Bayle quotes in the article "Manicheans":

> (...) this ancient opinion (...) can be traced to no source, but it carried a strong and almost indelible conviction, and is in circulation in many places among barbarians and Greeks alike (...) Nature brings, in this life of ours, many experiences in which both evil and good are commingled (...) as the result of two opposed principles and two antagonistic forces (...) For if it is the law of Nature that nothing comes into being without a cause, and if the good cannot provide a cause for evil, then it follows that Nature must have in herself the source and origin of evil, just as she contains the source and origin of good. The great majority and the wisest of men hold this opinion: they believe that there are two gods, rivals as it were, the one the Artificer of good and the other of evil.[47]

In the article "Zoroaster," Bayle again makes extensive use of Plutarch. He notes that, commenting on Plato's doctrine of the making of the world and of the two souls, in his treatise *On the Generation of the Soul in the Timaeus*, Plutarch

> makes us see that the origin of evil is not in an insensible, inanimate matter, which has no action, no qualities, and which can take all imaginable forms; but in a matter which moves, and which is united to a soul whose disorders cannot be entirely and fully corrected (DHC, Zoroastre, E).

---

a single cause, good or evil, which is the principle of all things together, because God is not the cause of any evil, and the harmony of this world is composed of opposites, as Heraclitus said (...)". Ibid: "Plutarch, in another book [*That it is Not Possible to Live Pleasurably According to the Doctrine of Epicurus*], says formally that God's nature only allows him to do good, and not to be angry with anyone, or to harm them. This author must therefore have been persuaded that the afflictions which so often torment men have another cause than God, and consequently that there are two Principles, one which only does good, the other which only does evil. I would add that the Persian philosophers, much more ancient than those of Egypt, constantly taught this doctrine." See also DHC, Périclès, K.

[44] DHC, Pauliciens, E: "The claim of Zoroaster, Plato, Plutarch, the Marcionites, the Manichaeans, and in general all those who admit a naturally good principle and a naturally evil principle, both eternal and independent, is that otherwise it is impossible to say by what means evil came into the world. You answer that it came through man; but how can this be, since, according to you, man is the work of an infinitely holy and infinitely powerful Being?". Cf. DHC, Zoroastre, E, where Bayle says that Plutarch rejected the "fables" invented by the "Magi" (i.e. the Zoroastrians), that is the ludicrous details of their explanations about the cosmic struggle between good and evil, but not the "foundation of their system", that is the idea of two rival principles. One should note that this was not the case with all Platonists; see for instance Alexander of Lycopolis' *Critique of the Doctrines of Manichaeus*, and, of course, Plotinus' treatise against the Gnostics (*Enneads* II.9).

[45] Plutarch, *On Isis and Osiris*, 369 E-F, tr. § 46, 113.

[46] Ibid., 370 D–F, tr. § 48, p. 117–119. See also Plutarch, *On the Generation of the Soul in the Timaeus*, 1026 B, 253–255, which adds the name of Parmenides. On dualism in early ancient Greek philosophy, see Miller 2011.

[47] Plutarch, *Isis and Osiris*, 369 B–E, tr. §§ 45–46, 109–111. Cf. DHC, Manichéens, C.

Then Bayle quotes extensively Plutarch's treatise, 1014 A to 1015 E, (DHC, Zoroastre, E) and after this long excerpt, which, he says, he had to put before his readers' eyes, he concludes with enthusiasm:

> This development of Plato's doctrine on the creation of the world, and on the origin of evil, is one of the most beautiful passages to be found in Plutarch; and although this doctrine is not true, it nevertheless deserves to be read with attention, and contains beautiful ideas, and conceptions sublime and of marvelous fruitfulness for those who know how to take advantage of the consequences (DHC, Zoroastre, E).

It is clear that the reason for Bayle's excitement is that, in the passage he quotes at length, Plutarch argues that evil is not simply due to the existence of matter, as if evil were unavoidable because, as Plotinus says, matter is deficient, shadowy, scattered, rebellious against form and order, etc. Since matter is privation, non-being, this thesis ultimately suggests that evil itself is mere nothingness, a privation of being and goodness, as Augustine contended in his turn. But this is a conclusion that Bayle vehemently rejects: the terrible reality of evil is all too obvious.[48] Consequently, Bayle is delighted by Plutarch's argument that evil requires an efficient cause. But matter, by itself devoid of any property or power, cannot play this role.[49] The cause of evil cannot be the "good artificer" either. Therefore, if evil does not come out of nothing, as the Stoics absurdly suppose, another cause must be found, an active cause, a full-fledged agent, equal to and opposed to the metaphysical principle of goodness (God, the monad, or the One), since evil is always and everywhere present.[50] For Plutarch, it can only be the Indefinite Dyad, operating through the irrational, maleficent world Soul.[51] This precosmic Soul (it exists before the Demiurge

---

[48] See for instance DHC, Euclide, B, et "Mémoire": OD IV, 180.

[49] Plutarch, *On the Generation of the Soul in the Timaeus*, 1014 E-F, tr. § 6, 189: "Those, however, who attribute to matter and not to the soul what in the *Timaeus* is called necessity and in the *Philebus* measurelessness and infinitude (…) what will they make of the fact that by Plato matter is said always to be amorphous and shapeless and devoid of all quality and potency of its own (…)?" (cf. Timaeus 50b 6–50c 2, 50d 7–50e 1, 50e 4–5, and 51a 4–7).

[50] Ibid., 1015 A-B, p. 191–192 (modified): "whence did these [disorders] come to be in things if the substrate was unqualified matter and so void of all causality, and the artificer good and so desirous of making all things resemble himself as far as possible, and there was no third besides these? For we are involved in the difficulties of the Stoics by bringing in evil without cause and process of generation out of what is non-existent, since, of things that do exist, neither what is good nor what is without quality is likely to have occasioned evil's being or coming to be." Ibid., 1050 B, tr. § 6, 193: Plato "did not as they [the Stoics] did by overlooking the third principle and potency, which is intermediate between matter and god, acquiesce in the most absurd of doctrines that makes the nature evils supervenient I know not how in a spontaneously accidental fashion."

[51] Ibid., 1015 E, tr. § 7, 197: "what Plato calls the cause of evil is the motion that moves matter and becomes divisible in the case of bodies, the disorder and irrational but not inanimate motion, which in the *Laws*, as has been said, he called soul contrary and adverse to the one that is beneficent." Cf. 1014 E, tr. § 6, 187: Plutarch observes that in the *Laws*, that soul is "openly called disorderly and maleficent soul" (cf. *Laws* 896b 5–898c 8, especially 896e 5–6, 897b 3–4, 897b 1, and 898c 4–5, but p. 187, note *f* to Plutarch's text, H. Cherniss comments that "in fact, the passages of the *Laws* envisage no such evil 'word soul'"). Cf. *On the Obsolescence of the Oracles*, 428 F–429 D. On Plutarch's doctrine, see (Thévenaz 1938) and (Dillon 2002).

makes the world) imparts disorderly motion (identified with evil) to matter. As a remedy, God (the Good Principle) "put a stop to [matter's] being disturbed by the mindless cause [the Soul]." He gave the Soul harmony, proportion, and number so as to make it stable.[52] However, the chaotic nature of the Soul cannot be completely eliminated, and thus evil remains present in the world. Plutarch goes on to speak of the constant tension between the irrational and rational parts of the Soul, which alternately take over, and concludes:

> Thus many considerations make it plain to us that the Soul is not god's work entirely but that with the portion of evil inherent in her she has been arranged by god (...).[53]

This is the crucial point for Bayle: as Plutarch shows, from a purely philosophical point of view, evil is not supervenient, that is, it must be present originally in a principle. Bayle sums up the whole argument: "Where did evil come from, if the principle of evil is not eternal?", since it did not come from the principle of goodness nor from mere matter (DHC, Zoroaster, F).

Thus, when Bayle specifically needed a non-Christian antagonist to the unitarist conception, Plutarch could have been an alternative to Zoroaster or the Manicheans.[54] Bayle knows his vast body of work very well and refers to him often— the *Dictionary* contains 1087 occurrences of the name "Plutarch," and the rest of Bayle's works 139. There are enough arguments in Plutarch's writings to make him, in the debates organized by Bayle, the eminent representative of an irrefutable dualistic point of view. As a matter of fact, he would have been a more formidable opponent of unitarism than the Zoroastrians, the Manicheans, or the Gnostics, who left not philosophical works but myths full of inconsistencies and, according to Bayle, were not skilled debaters. On the contrary, Bayle admires the way Plutarch refutes the Stoics on the question of evil. In addition to the passages from *On Isis and Osiris* and *On the Generation of the Soul* we have just read, Bayle quotes liberally from Plutarch's *Against the Stoics on Common Notions*. He highly approves of Plutarch's rebuttal of the Stoic claim that the existence of evil things was necessary for the good of the cosmos, just as the existence of vice is necessary to the existence of virtue (Pauliciens, G). He also parallels the discussion between Plutarch and the Stoics with that between Melissus and Zoroaster in the article "Manicheans."[55] In the arti-

---

[52] Ibid., 1015 E, tr. § 7, 197.

[53] Ibid., 1027 A, tr. § 28, 261. Cf. *Isis and Osiris*, 371 A, tr. § 49, 121: "Yet it is impossible for the bad to be completely eradicated, since it is innate, in large amount, in the body and likewise in the soul of the Universe, and is always fighting a hard fight against the better."

[54] Indeed, Bayle names Plutarch together with Zoroaster in that context: "For if we were dealing with Zoroaster or Plutarch, it would be a different matter" (DHC, Pauliciens, E) by contrast with a Gnostic or other heretic interlocutor who accepts part of the biblical Revelation and might be led to recognize the divine inspiration of the Old Testament, in which case one can use the reasoning based on the fact of Adam's and Eve's fall as proof that it was possible for the good principle to let evil happen (ibid.)—which is the solution Bayle recommends in this kind of discussion (see below, section 8).

[55] Pauliciens, G: "Melissus and Parmenides weren't the only ones who were troubled by these difficulties; the Stoics also found them highly embarrassing. Without denying that there were many

cle "Chrysippus," Bayle notes that Plutarch accuses the Stoic philosopher of making God the author of sin and he confirms the charge.[56] Then, following Plutarch' steps, Bayle refutes the solution attempted by Chrysippus, who, like Melissus and the Christian theologians, blames human freedom for the existence of evil. Chrysippus makes a distinction between main causes and auxiliary causes, and claims that God, through fate, only prepares auxiliary causes, that is, occasions for our actions, but that the main cause of our actions is the disposition of our own soul. We act according to what we are, that is, freely. He famously illustrates this idea with a cylinder: it may need an external push to start moving (auxiliary cause), but if it then rolls (unlike a cube, for instance), this is due to no cause other than its shape (principal cause). Bayle objects that the shape of the cylinder itself has a cause: a craftsman. The application of the objection to the matter at hand is obvious.[57]

As a consequence, Plutarch could have served perfectly well as a protagonist for what Bayle intended to show in the article "Manicheans" and in other similar places. Since there were ample resources in the Western philosophical heritage, we must therefore return to our question: if not for a touch of exoticism, why did Bayle use representatives of dualistic Eastern religions?

7.6 The Accusation of "Manicheism" in Anti-protestant Controversies

The answer to the question above is given by the context of Bayle's writings. In fact, Bayle was not the first author to conjure the ghost of Manicheism in modern times. It was customary in the religious controversies of the day to accuse one's

---

gods, the Stoics reduced them all to Jupiter, as the sovereign dispenser of events. It was to him that they attributed Providence, and they recognized him as an infinitely good and infinitely prudent Being. This is the basis of Plutarch's objections to them, taken from the misery of the human race. 'There is no wise man,' he says, 'nor was there ever on earth, and on the contrary there are innumerable millions of unhappy men in dire straits, under the oversight and rule of Jupiter, whose government and administration is very good. And what could be more contrary to common sense, than to say that Jupiter governing sovereignly well, we are sovereignly unhappy?'".

[56] DHC, Chrysippus, H (with cross-reference to DHC, Pauliciens, G, for more details on Plutarch's refutation): "According to Chrysippus, God was the Soul of the World, and (...) the world was the universal extension of this soul, and (...) Jupiter was the eternal Law, the fatal Necessity, the immutable truth of all future things. The necessary and inevitable consequence of this is that the soul of man is a portion of God, and that all his actions have no other cause than God himself."

[57] Ibid:"(...) For, in order to make an accurate comparison with the cylinder, we must compare destiny, not with the first person who pushes it, but with the carpenter who made it, and who then pushes it with his foot. The fact that the cylinder rolls for such a long time comes from its shape, but because the carpenter gave it this shape, the necessary cause of lasting movement, he is the real cause of the duration of this movement. All the difference between a cube that doesn't roll, and a cylinder that does, all the consequences, all the regularities or irregularities of the rest of the one, and the continued movement of the other, must be attributed to the workman who gave these two bodies the shape from which they necessarily result. Everyone can apply this to human souls."

adversaries of falling into Manicheism—not in the sense of adopting real diarchism, but in the sense of making the Christian God the ultimate cause of sin, as if he were the Evil Principle of the Manicheans. Bayle makes this point perfectly clear.

> The style of the orthodox does not change on this point. From time immemorial it has settled on the following usage: It has been established from time immemorial that to be a Manichaean and to make God the author of sin are two expressions which mean the same thing; and when a Christian sect accuses others of making God the author of sin, it never fails to impute Manichaeism to them in this respect (DHC, Pauliciens, I).[58]

Such an accusation was commonly hurled at the Protestants by the Catholics. As early as the sixteenth century, Lutheranism and Calvinism were denounced by the Catholics as a revival of Manicheism because they negate human freedom of indifference in the present life and assert that without God's grace humans can only sin, as if they were under the rule of the Evil Principle.[59] Thus, Erasmus, in his *De libero arbitrio*, contrasts Luther's thesis on the bondage of the will with the broad consensus of the tradition about the reality of free choice, and he places him in the company of the Manicheans and of Wyclif.[60] The English Catholic bishop John Fisher (1469–1535), an early opponent of Luther and a future victim of Henry VIII, does the same,[61] as does the Italian Capuchin Lawrence of Brindisi (Giulio Cesare Russo, 1559–1619).[62] In the next century, Jansenius, in his *Augustinus*, balances his denunciation of the modern Pelagians that the Jesuits are with a denunciation of the modern Manicheans that the Calvinists are to his mind. Just as Augustine fought the two opposite errors of Pelagianism and Manicheism, between which a middle way must be found, so Molinism and Calvinism must be likewise repelled as two symmetrical

---

[58] This polemical use of the term "Manichean" has been noted by Hickson 2013, 56–63.

[59] It is true that the Protestants sometimes showed some sympathy for the Cathars, but it was not inasmuch as they were dualists. Rather, it was inasmuch the Cathars had opposed the Roman Church, denouncing the clergy's corruption and rejecting transubstantiation, the invocation of saints, the veneration of relics and of images, and the prayers for the dead. For the same reasons, the Protestants were sympathetic to Pierre Valdo and his disciples. But they vehemently rejected the accusation of diarchism. See Walther 1968; Carbonnier 1955; Krumenacker 2006.

[60] Erasmus, *De Libero Arbitrio*, I b 2, tr. (ed. J. von Walter), 12.

[61] Fisher, *Assertionis Lutheranae confutatio*, Articulus XXXVI, p. 196: "*Liberum arbitrium post peccatum res est de solo titulo, et dum facit quod in se est, peccat mortaliter*. Manichaei primùm duas in homine mentes faciunt: alteram quam dicunt malam, et substantiam quandam mentis adversae, nimirum eandem quam et nos vocamus concupiscentiam carnis. Alteram aiunt esse mentem bonam, quam et nos vocamus Spiritum. Has duas afferunt in hominibus mutuum habere conflictum, cùm caro concupiscit adversus spiritum, et spiritus adversus carnem. Peccata verò quae fiunt, ex mala mente proficisci contendunt, nec per liberum arbitrium vitari posse quovis pacto."

[62] Laurentius a Brundusio, *Lutheranismi hypotyposis*, pars I, sectio 5, dissertatio 3, cap. 7, 255: "Summam autem istam blasphemiam Lutherus in libro De Servo Arbitrio in Erasmum Roterodamum satanico spiritu eructavit; (…) Attendit Lutherus spiritui Manichaeorum, qui, ut S. Hieronymus scribit, hominum damnabant naturam et liberum auferebant arbitrium." See references to the *Catalogus haereticorum omnium* by Bernhard of Luxemburg, and to the Jesuit Jacob Gretser in Walther 1968, 184–185.

errors.⁶³ Following this example, Blaise Pascal, in his *Écrits sur la Grâce* (1656), writes that "the Manicheans were the Lutherans of their time, just as the Lutherans are the Manicheans of ours."⁶⁴ As Richard Popkin has noted, Bossuet, in what became a major piece of seventeenth–century anti-Protestant controversy, his *Histoire des Variations des Églises Protestantes* (1688), contended that the Cathar Manicheans were the ancestors of the Protestants (1967, 32). Similarly, the Jesuit Maimbourg, a recurrent target of Bayle's indignation, has the reformed confession descend in a straight line from the Waldenses, whom he does not distinguish from the Cathars.⁶⁵ Let us note that these Catholic charges of Manicheism had implications for a question dear to Bayle: that of tolerance. Indeed, like Augustine's policy towards the Donatists, the extermination of the Cathars was taken by some as a model for dealing with Protestants.⁶⁶

Another reason for the Catholics' accusation of Manicheism was the Calvinist doctrine of predestination and the resulting image of God it presents. The fact that God withholds his grace from some people (the vast majority of them, in fact) seems to make him the cause of their sinning (all the more so when, as in the theory of double predestination, he predestines them to damnation). The Huguenot minister Jean Daillé (1594–1670) summarizes the attacks of the Jesuit Jean Adam (1608–1684) as follows:

> [according to Adam] we say that this same sovereign God and Lord, whom you and we worship, is a deceitful, cruel and inhuman God: a God without justice, without reason and without goodness (...) And that when the Marcionites and Manichaeans decided to make a second God the author of all evils, they at least worshipped another who gave all good things, whereas ours is worse than men.⁶⁷

---

⁶³ Jansenius, *Augustinus* (Jacob Zegers 1640), Liber sextus. De gratia Christi salvatoris qui est primus de libero arbitrio, praefatio, p. 1137: "Dicunt illi Manichaei quibus modo non communicamus, id est, toti isti, cum quibus dissentimus, quia primi hominis peccato, id est, Adae liberum arbitrium perierit. (...) (*Lib.* 1. *ad Bonif. c.* 2)." Ibid., p. 1138: "Possunt duo errores inter se esse contrarii, sed ambo sunt detestandi, quia sunt ambo contrarii veritati. Nam si propterea diligendi sunt Pelagiani quia oderunt Manichaeos, diligendi sunt Manichaei quia oderunt Pelagianos (Lib. 2. *ad Bonif.* c. 2)".

⁶⁴ Pascal, *Quatrième Écrit*, 340. Quoted by Popkin 1967, 33.

⁶⁵ Maimbourg, S. J., *Histoire du Calvinisme*, 1682, p. 69. Quoted by Carbonnier 1955, 82.

⁶⁶ See (Carbonnier 1955, 84 n. 44). Cf. Louis de La Valette, *Parallèle de l'Hérésie des Albigeois et de celle du Calvinisme, dans lequel on fait voir que Louis le Grand n'a rien fait qui n'eût été pratiqué par S. Louis* (1686), 586 (cité par Bayle, CP: OD III, 404).

⁶⁷ Jean Daillé, *Réplique de Jean Daillé aux deux livres que Messieurs Adam et Cottiby ont publié contre lui*, 2ᵉ éd., part II, chap. 1, p. 2–3. In Jean Adam, *Réponse à la lettre de M. Daillé*, p. 139–141, Adam uses a language that is closely similar to Bayle's Manicheans' conclusions: "They make him out to be a God carried away with anger and fury against men, who resolves to lead them to perdition and damnation (...). They say that he is a deceitful God, who has two wills: one public, by which he declares that he wants to save everyone; and the other secret, by which he pushes into impiety those he does not love, in order to find a pretext to punish them. They also say that he is a cruel God, who gave the old Law only to make men more criminal, by depriving them of the power to fulfill it; and that he is an inhuman master, who commands impossible things from his servants (...) And when the Catholic Doctors (...) want to relieve God of the horrible slander with which they accuse his goodness as if he were complicit in all crimes, by way of the distinction they have found

The accusation of Manicheism was also common in inter-Protestant controversies. It is an accusation that the arch-Calvinist Pierre Jurieu leveled at Luther and that the Lutherans leveled at the Calvinists.[68] After the Lutheran theologian Mathias Flacius Illyricus (1520–1575) maintained that the original sin had corrupted so deeply the human soul that sin had become "the substance of man," he was charged with Manicheism by his opponent, Viktorin Strigel, a fellow Lutheran but one of the followers of Melanchthon called "Synergists", who thought that free-will had not been totally destroyed.[69] Within the Calvinist camp, a number of theologians broke with the doctrine of predestination. In the Netherlands, at the end of the sixteenth century, Jacob Arminius started a dissenting current (named Arminianism, but his followers were also called "Remonstrants") that clashed with Calvinist orthodoxy. It was condemned in 1618–19 at the synod of Dordrecht, but remained very much alive and influential. Similarly, in France, Reformed theologians such as John Cameron, Moïse Amyraut, and Claude Pajon, who sought to soften the rigorism of the predestination theory and hypothesized a universal grace given to all mankind, got in trouble with Calvinist hardliners (Du Moulin, Daillé, Claude, and the like), who admitted of only individual grace bestowed on those who are predestined to receive it (DHC, Amyraut, main text). As for the Socinians (named after Lelio and Fausto Sozzini), whose characteristic was to critically examine Revelation in the light of reason (thus coming to reject the Trinity dogma), they were generally reviled by the other Reformed theologians, but they had a common ground with Arminians and Pajonists in finding that Calvinist orthodoxy incriminates God as being ultimately the cause of sin. Thus, Charles Le Cène— an Arminian, or even, according to his enemies, a Socinian—accused Augustine (whose doctrine the Calvinist "*prédestinateurs*" claim to follow) of having remained in effect a Manichean, and even of being worse than the Manicheans:

> In truth, Sir, I would rather adopt the doctrine of the Manicheans on this matter than that of St. Augustine. Even the Manicheans do not push the matter to the limits as strongly as St. Augustine when he broke up with Manicheism; they at least admit that among men, there were a very large number who had a good nature; St. Augustine, instead, believed them all generally and equally corrupted by the sin of Adam (...) there is nothing that is further from the creed of the whole Church, before St. Augustine drew his doctrine from the school of the Manicheans (....).[70]

---

between the things God wills and does, and those he allows without willing them, they reply that this distinction is frivolous and idle, and they strongly maintain that God's supreme will precedes all the actions of his government and that it is the necessary cause of all the crimes that occur in the world."

[68] Quoted by Bayle, DHC, Pauliciens, F: "After reporting Luther's sentiments, he says, *Hæc omnia abdicamus et horremus ut religionem omnem pessundantia et Manicheismum spirantia*. Petrus Jurius, de Pace inter Protestantes ineunda, *pag.* 214. See Mr. de Meaux in the *Addition* to *Histoire des Variations*."

[69] See DHC, Illyricus (Mathias Flacius), C, and DHC, Synergists. Let us note that Bayle approves, if not of its formulation, at least of the gist of Illyricus's view (which is close to Calvin's) that sinning is not epiphenomenal ("accidental", in Aristotelian parlance) but substantial to the soul.

[70] Charles Le Cène, *Entretiens sur diverses matières de théologie*, Première Partie, 115–116.

As we can see, the charge of "Manicheism" was a common slur,[71] used by all parties in sixteenth– and seventeenth–century religious disputes, but particularly flung at the Calvinists for seemingly making God responsible for sin. This polemical use of the term suggests that the "Manicheans," in Bayle's *Dictionary*, do not suddenly appear out of the blue, emerging only from Bayle's fertile imagination, but that Bayle had a specific purpose in giving them a prominent function. As I will show in the next section, his purpose was to defend Calvin's doctrine against the slanderous charge of Manicheism by proving that those who want to explain away evil on the basis of free will fare no better with regard to God's responsibility.

7.7 Bayle's Defensive Strategy

Indeed, this historical background of religious controversies best accounts for Bayle's complex strategy in the *Dictionnaire*, as he fights on at least three fronts: against the Catholic anti-Protestant controversialists, against his fellow Protestants who were leaning toward Arminianism or Socinianism, and against Pierre Jurieu regarding toleration. I am in complete agreement with the interpretation that Dmitri Levitin has recently formulated:

> if we place the famous Manichean articles of the Dictionnaire – 'Manichéens', 'Marcionites', and 'Pauliciens' – in their proper context, we shall find that they were neither an expression of fideism (or at least, that they were no more fideistic than the theological opinions held by Bayle's Reformed counterparts) nor atheism. Rather, they were designed to serve two polemical purposes. First, they were designed to defend the rationality of believing in Reformed predestinarian dogma. As we shall see, Bayle followed in the footsteps of several Reformed theologians – many of whom he knew personally – in arguing that while the problem of evil was unsolvable, the most rational path to adopt in the face of the irreconcilability of divine power and human sin was to surrender oneself to a predestinarian theology that at least had the benefit of best respecting the divine excellence. (…) The second polemical aim of the Manichean articles was to contribute to Bayle's case for toleration (Levitin 2022, 308–309).[72]

---

[71] Just as the term "atheist," with which, as Febvre 2003, 126–138, has noted, Renaissance authors were all too happy to pepper their polemics.

[72] See ibid., 241: "It is here that we find the origins of the famous Manichean articles of the Dictionnaire, (…). What previous interpreters have failed to recognise is that those articles had a double polemical function. On the one hand, they sought to demonstrate against Jurieu that the revealed mysteries were always beyond the grasp of human reasoning, and so were by definition subject to invincible ignorance. On the other, they sought, with Jurieu, to defend Reformed predestinarianism by showing that it was no more irrational than any 'relaxed' form of predestinarian dogma (i.e. as upheld by Molinists, Arminians, and Pajonists) and that it was in fact more rational to stick with the 'rigid' position of the Reformed." See also Hickson 2013, who connects the Manichean articles to the question of toleration and correctly sees that Bayle's reassessment of the Manichean position is in fact a defense of Calvinism.

Having looked, in the previous section, at other modern texts besides Bayle's, texts in which the Manicheans loom too, we must come to the same conclusion.[73] Bayle is well aware that the Calvinist doctrine of predestination makes God look like an unjust and cruel tyrant.[74] But in response to the attacks on it coming from various quarters, he shows that the opponents' solution to the problem of evil, which in all its variants rests on the responsibility of human free will, does not actually solve the problem and exposes them no less than the Calvinists to the charge of making God the author of sin, since God is ultimately responsible for giving humans free-will and letting them make a bad use of it. In short, the reproach of Manicheism addressed to Calvinism can be turned against the attackers, whether they be Catholics, Arminians, or Socinians. If the specter of Manicheism is recalled from its grave, it will haunt all Christian denominations equally. By the lights of reason, every Christian theology, at one point or another, seems to render God responsible for the existence of evil, as Bayle's diarchists are in charge of demonstrating. Consequently, Bayle's point is (1) that the Calvinist position does not do worse than any other, and (2) that its adversaries should drop the charge of Manicheism, to which they themselves are exposed. Bayle's intention to turn the tables specifically with regard to the accusation of Manicheism explains why, when he needs a diarchist, he uses Manicheans (or, equivalently, Zoroastrians, since, as we have seen, he makes no distinction between the two) as unlikely protagonists in the debate he stages, rather than, say, Plutarch.

This strategy becomes quite clear when one pays attention to the composition of remark F of the article "Paulicians." At the beginning of his long tirade against the sterility of the disputes about predestination, Bayle remarks: "Since Luther and Calvin came into the world, I don't think a year has passed in which they haven't been accused of making God the author of sin" (DHC, Pauliciens, F). Then, in the next column, the Manichean dismisses all the Christian sects alike and tells *all of them*, even the Molinists or the Socinians (who grant the greatest role to our free will), that they are Manichean without knowing it or acknowledging it, because, just like the diarchists, they posit two principles, one the root of goodness, the other the root of evil.[75] Moreover, they do it in the most monstrous fashion, since they place

---

[73] Let us open here a methodological parenthesis: interpretations of Bayle have generally suffered from too narrow a focus on his texts and on his immediate circle of friends and enemies. Even a whole book dedicated to both Bayle and Jurieu, such as Van der Lugt 2016, does not manage to clarify the respective positions of the two authors and perpetuates stereotypes and intimations on Bayle's intentions. It is one of the great merits of Levitin's book to have taken a very broad view of Bayle's historical context, which enabled it to rectify the perspectives on Bayle's mindset and work.

[74] This does not mean that he concedes it is the case. As we'll see below, he has a solution that negates this claim. Let us just note here that Bayle gives as a model the fairness of Melanchthon, who, while he thought that Calvin's doctrine made God the author of sin, was nevertheless equitable enough to acknowledge that Calvin would have vehemently rejected that conclusion (DHC, Melanchthon, B).

[75] DHC, Pauliciens, F: "If you examine your system carefully, you will recognize that, as well as I do, you admit two principles, one of good, the other of evil."

both principles in one and the same God—hence the Manichean's invitation, as we saw earlier, to abandon a "less reasonable Manicheism" for a "more reasonable" one by admitting that the two principles are independent each from the other.[76] Thus it is evident that Bayle responds in kind to the recurring accusation of Manicheism against Luther and Calvin, saying in sum: You too, who want to lay the blame on human free will, in fact make God the author of sin.

This way of responding is one of Bayle's favorite dialectical devices: It is a *rétorsion* (an *anticategoria* or *metastasis, accusatio adversa, translatio in adversarium* in the nomenclature of ancient rhetoric), that is to say, a counter-accusation, a reply made to an argument by redirecting it against the attackers and showing that they are affected by the very objection they have raised. The outcome is that the defenders don't have to abandon their position, since the opponents' position suffers from the same flaw and is therefore no better. As a result, in the matter at hand Bayle has no reason to abandon the Calvinist predestination theory, since all other theologies likewise make God responsible for the existence of evil.[77] Bayle wryly refers to his *frère ennemi*, Pierre Jurieu, who drew exactly the same conclusion, and with whom he is in fact in agreement on this count:

> I refer you to a professor of theology who is still alive, and who has shown, clear as day, that neither the method of the Scotists, nor that of the Molinists, nor that of the Remonstrants, nor that of the Universalists, nor that of the Pajonists, nor that of Father Malebranche, nor that of the Lutherans, nor that of the Socinians, are able to withstand the objections of those who impute to God the introduction of sin, or who claim that it is not compatible with his goodness, nor with his holiness, nor with his justice; so that this professor, finding no better elsewhere, abides by St. Augustine's hypothesis, which is the same with that of Luther and Calvin, and of the Thomists and the Jansenists (…). (DHC, Pauliciens, F)

This strategy is also the reason why Bayle places so much emphasis on the question of free will in these discussions. As we saw earlier, when he decides to supplement the shortcomings of Manichean answers to the criticisms of various Fathers of the Church, he vigorously challenges the latter's explanation. Several times, Bayle underlines the point that the current disputes among Christians about free-will would have made it easier for a Manichean to refute unitarism:

> The disputes that have arisen in the West among Christians since the Reformation have so clearly shown that one doesn't know where to start when one wants to resolve the difficulties

---

[76] See above, section 7.3, at fn. 28.

[77] EMT: OD IV, 7: "(…) these objections to the system of absolute predestination are established with much more skill in the books of the Jesuits, the Lutherans and the Remonstrants than in the *Critical Dictionary*. I observe a similar thing with regard to the objections to the systems of these three sects. The Thomists, the Jansenists and the Calvinists have struck them down with far greater force than Mr. Bayle. There is nothing more unjust than to find it wrong for those who have examined these major disputes to acknowledge ingenuously that no party can satisfy the difficulties objected to it. The Reformers admit that we must stop at the edge of this abyss, and do not boast of satisfying reason; but they make it clear that their opponents' objections prove too much; they retort them all, and challenge their antagonists to parry the retort."

about the origin of evil, that a Manichaean today would be more formidable than in the past, for he would refute us all, each one by the others.[78]

That is to say, these disputes have left a stockpile of arguments *pro et contra* that a modern Manichean could use to refute any Christian system that tries to account for the existence of evil without blaming God for it: the Molinists have been refuted by the Jansenists, the Arminians by the Calvinists, and vice-versa.[79] The Manichean would simply have to conclude that there is no explanation capable of exonerating God other than the existence of an evil principle.

Of course, Bayle is not content with simply stating that Calvinist theology is no worse than other theologies with respect to God's responsibility. The counter-accusation is only designed to neutralize the polemical charge of "Manicheism," not to acknowledge that indeed the Calvinist doctrine renders God responsible for our sins. Beyond scoring that point, Bayle wants to affirm that God *is not* the cause of evil. The solution he offers is based on both reason and Revelation, as I will explain at the end of the next section, after talking about the place of Revelation in the humble sort of theology Bayle calls for.

7.8 Bayle's Irenicism

In fact, in Bayle's eyes as in Jurieu's, the Calvinist position is better than the others in several respects. In particular, it is closer to the Scriptures and acknowledges God's absolute sovereignty. That is what Bayle suggests when he lends these words

---

[78] DHC, Pauliciens, F. See also ibid., main text: "This hypothesis of the two principles would apparently have made more progress, if it had been less crudely detailed, and if it had not been accompanied by several odious practices, or if there had then been as many disputes as there are today about Predestination, in which Christians accuse each other of either making God the author of sin, or of taking away from Him the government of the world." Also, DHC, Manichéens, D: the Christians "have more difficulty than the Pagans in clearing up these difficulties by means of reason, because they have disputes among themselves about liberty, in which the aggressor always seems to be the stronger (…)."

[79] DHC, Pauliciens, F: "You have exhausted, he would tell us, all the powers of your mind. You invented Middle Science, as a sort of *deus ex machina* who comes and sort out your chaos. This invention is chimerical; it is incomprehensible that God could see the future elsewhere than in his decrees, or in the necessity of causes. It is no less incomprehensible according to metaphysics than it is according to ethics, that, being goodness and holiness itself, God should be the author of sin. I refer you to the Jansenists: see how they strike down your Middle Science, both by direct proofs and by the retorsion of your arguments; for it does not prevent all sins, and all the misfortunes of man, from coming from God's free choice (...) As for absolute decrees, the certain source of prescience, see, I beg you, how the Molinists and Remonstrants combat them. Here is a theologian as resolute as Bartole, who confesses almost with a tear in his eye that there is no one who is more bothered than he is by the difficulties of these decrees, and that he remains in this state only because, having wanted to transport himself into the methods of relaxation, he still finds himself overwhelmed by these same burdens."

to a (fictional?) Catholic priest, who is in the same predicament and seems to envy the *Sola Scriptura* principle of the Protestants:

> (…) we find ourselves in the middle of an intersection, about which a learned priest in Paris said the following not long ago: I have four roads around me, that of the Calvinists, that of the Jansenists, that of the Thomists, and that of the Molinists. I know which one not to take, but not which one to take. *Quem fugiam habeo, quem sequar non habeo*: the first road is contrary to the Council of Trent, the second to the Popes' Constitutions, the third to Reason, and the fourth to St. Paul. Those who are not Roman Catholics can get out of this predicament more easily by preferring the authority of St. Paul to that of the Popes and the Councils (DHC, Pauliciens, N).

However, Bayle refrains from harping on the superiority of Calvinist theology. His intent is not to draw out that kind of discussions, but, on the contrary, to quell them, as they do more harm than good to Christianity. The last resort responsibility regarding evil is one of those questions that, he says, it would be better to let rest like the old markers that delimit the lands people have inherited (DHC, Stancarus, H). All these deadlocked debates show that controversies are pointless. Bayle shares Averroes's condemnation of theologians who indulge in unrestrained interpretations of Revelation and recklessly spread these interpretations in the community, disturbing the faith of ordinary believers, and stirring up quarrels and schisms:

> It is because of the interpretations, and because of the opinion that these should, from the point of view of the revealed Law, be exposed to everyone, that the sects of Islam arose, which came to the point of accusing each other of infidelity or blamable innovation [...] They thereby precipitated people into hatred, mutual abhorrence and wars, tore the Revelation to pieces and completely divided men.[80]

Similarly, Bayle thinks that professional theologians are guilty of disturbing the peace when they discuss questions that are beyond human comprehension and can never be the subject of a consensus, leading straight to that disease of the divines, the *rabies theologorum*.[81] While philosophers should be allowed to philosophize, i.e. to think according to their own standards and draw their conclusions (which is what Bayle claims for himself), on the other hand one should refrain from raising eristical questions about Scripture and let ordinary believers simply accept its obvious meaning, without trying to split hairs. The best move, Bayle suggests, would be to close theology schools.

Given the centrality of the question of free will in Bayle's strategy, the inclusion, in one of the "Manichean articles", the article "Marcionites" (end of rem. F, n. 52),

---

[80] Averroès, *Discours décisif*, § 64, p. 165. See Solère 2024.

[81] Cf. DHC, Amyraut, E: "(…) argue only as much as you can do it without disturbing public peace, and keep quiet as soon as events show you are dividing families, or forming two parties. Don't end up awakening a thousand bad passions, which must be kept chained like so many ferocious beasts; and woe betide you, if you cause them to break their fetters." DHC, Amyraut, F: "Where did the initial uproar against this system come from? How did the same doctrine come to be seen first as a monster, and then as an innocent thing? Shouldn't we recognize in this the trace of the original sin, and the influence of a thousand dark passions, which must finally produce, if we are among the predestined, a salutary and mortifying humiliation? What's worse, one doesn't profit from the past: each generation brings the same symptoms, sometimes greater, sometimes lesser."

of a reference to Bernardo Ochino's *Labyrinthi, hoc est, de libero aut servo arbitrio, de divina praenotione, destinatione, et libertate disputatio*, is significant, as that book calls for restraint and for a confession of ignorance regarding the insolvable problem of free will and predestination. Ochino (1487–1564) was an Italian Franciscan, then Capuchin, who converted to the Reformation in 1542. In the work cited by Bayle, he argues for and against free will in relation to divine predestination, showing that no certainty can be achieved. In his final chapter, probably echoing Nicholas of Cusa, he concludes that we must be content with "learned ignorance" (Ochino, *Labyrinthi*, 245). Through Revelation, God has given us everything we need to know in view of salvation, but he has provided no clue as to how free will is compatible with predestination. The resolution of this problem is therefore irrelevant to salvation. We must recognize that it is beyond our reach, and therefore we must refrain from discussing it. Bayle devotes an entire article to Ochino in the *Dictionnaire*, highly praising the *Labyrinthi* in remark P:

> They seemed to me to be the work of a man with a very clear and penetrating mind. In it, Ochino shows with great force that those who maintain that man acts freely get bogged down in four great difficulties, and that those who hold that man's actions are determined by necessity fall into four other great embarrassments; so much so that he delineates eight Labyrinths, four against free will, and four against necessity. He turns on every conceivable side to try to find a way out, and finding none, he concludes each time with a fervent prayer to God for deliverance from these abysses. Nevertheless, in the remainder of the work, he endeavors to point out a way out of this prison; but he concludes that the only way out is to say, like Socrates: unum scio quod nihil scio. We must be silent, he says, and judge that God requires from us neither affirmation nor negation on matters of this nature.[82]

The objections of Molina, Arminius, Cameron, Pajon, Sozzini, and their followers against the predestination so clearly affirmed by the Bible were fruitless, for their escape route through free-will is handily blocked by the Manichean objections orchestrated by Bayle. They should have simply accepted Revelation as it is, instead of occasioning controversies.[83] This is the meaning of Bayle's "fideism," as Dmitri

---

[82] DHC, Ochin, P. Also, DHC, Pauliciens, M: "the natural answer to the question, *Why did God allow man to sin?* is to say *I don't know, I only believe that He had reasons for it that are worthy of his infinite wisdom, but that are incomprehensible to me*. With this answer, you will stop the most obstinate disputants in their tracks; for if they wish to continue to argue, you will let them speak alone, and they will soon fall silent. If you were to enter the fray with them, and commit to prove them that the inviolable privileges of free-will were the real reason that led God to allow men to sin, you would be obliged to satisfy them on the objections they would make, and I don't know how you'd be able to do away with them; for they could oppose you two things which seem very obvious to our reason." DHC, IIe Ecl. (IV, 635): "It is therefore clear that the dogma of Adam's sin, with all that depends on it, is, among all the mysteries inconceivable to our reason and inexplicable according to its maxims, the one that most necessarily requires submission to revealed truth, notwithstanding all the oppositions of philosophical truth."

[83] See Levitin 2022, 372: Arminius "had hubristically abandoned Pauline self-restraint so as to enquire into the mystery of predestination and attack the biblical-Reformed view. But he had done it so unsuccessfully that his followers also had to make 'a sincere admission of the infirmity of our minds, or by the consideration of the incomprehensible Infinity of God' (…). 'Was this worth the effort of contradicting Calvin?', asked Bayle (DHC, Arminius, E), clearly presupposing the answer." Cf. DHC, Amyraut, E: "(…) What is the use and *cui bono* of these disputes? Are there no

Levitin has pointed out. Following in the footsteps of Calvinist theologians such as François Turretin and Jacques Abbadie, he reaffirms the need for a modest theology that limits itself to what Revelation states, instead of trying to explain it when it is incomprehensible:

> According to them [Turretin and Abbadie], not only had philosophical speculation led to interminable *odium theologicum*, but, in the final reckoning, the 'rationalist' arguments of the Pelagians no more solved the problem of reconciling divine power and prescience and human free will than did Reformed predestinarianism, while lessening divine omnipotence (Levitin 2022, 355).

Nevertheless, as I said above Bayle does affirm that, despite the appearances and all the Manichean arguments, God *is not* the cause of evil. The response he offers is in part faith-based. Indeed, he repeats often that the only possibility, for a Christian debating with a diarchist, is to retreat into the safe refuge of Revelation while confessing that reason is unable to solve the Manicheans' objections.[84] However, Bayle's position is not "blind fideism" as it is sometimes called. It rests, for another part, on the rational, axiomatic notion (a "common notion") that God always acts rightly ("whatever God does is done well") (Solère 2023a, b). Faith only intervenes to provide a content to which that principle can be applied. This content consists of the factual truths disclosed by the Bible, which cannot be known except through it, such as that God let Adam and Eve sin. From this, which faith holds to be a fact, one can reason from actuality to potentiality, or fact to possibility, and conclude that it was possible that a good God allow evil to be introduced in a world that is entirely

---

difficulties left, provided we use Cameron's hypothesis? Is it not true, on the contrary, that no remedy has ever been as palliative as this one? We need something else to satisfy reason; and if you don't go any further, you might as well not budge from where you were: remain at rest in particularism. I admit that Universalism has some advantages and responds better to certain objections. But is this enough to balance so many spiritual crimes that factions drag after them, so many mean suspicions, so many sinister interpretations, so many false imputations, so many hatreds, so many insults, so many libels, so many other disorders, which come in droves as a result of such a theological conflict?". DHC, Arminius, E: St Paul recognized in the problem of predestination "an incomprehensibility which must stop all disputes, and impose a profound silence on our reason. (...) All Christians must find therein a definitive ruling pronounced as a last resort and without appeal, concerning the disputes about grace; or rather, they must learn by this conduct of St. Paul never to dispute about Predestination, and to oppose at once this barricade to all the subtleties of the human mind, whether they arise of their own accord while we are meditating on this great subject, or whether another man proposes them to us. The shortest and best way is to first oppose this strong dike to the floods of reasoning (...)."

[84] DHC, Rufin, C: "(...) let us stand here immobile and unshakeable, putting our finger over our mouths and quieting our feeble lights, convinced that in these matters the best use of reason is not to reason at all." DHC, Pauliciens, F (III, 629b): "It is more useful than one might think to humble man's reason, by showing him how forcefully the most insane heresies, such as those of the Manicheans, deceive his lights, in order to confuse the most essential truths. This should teach the Socinians, who want reason to be the rule of faith, that they are throwing themselves into a path of error, which is only apt to lead them from degree to degree to deny everything, or to doubt everything, and that they are committing themselves to being beaten by the most execrable people. So what's to be done? One must captivate one's understanding under the obedience of faith, and never argue about certain things."

under his control, contrary to the Manicheans' claim.[85] Admittedly, our reason must confess that it cannot see why a good God would do such a thing and *how* that was possible, given the self-evident moral principles that God's decision seems to contradict.[86] Reason, however, must bow before the facts (as they are for the faithful), because it is limited and cannot understand everything.[87] It can nonetheless have confidence in the higher axiom, "whatever God does is right," even though it does not enable us to see how God's action, in the case in point, was actually right.[88] This particular combination of reason and faith is not tantamount to trying to explain God's conduct, as hubristic theologies do.

Furthermore, a self-restraining theology, as close to the Scriptures as possible, serves Bayle's purpose of ending the controversies that tear Christianity apart, for the sake of a tolerant society. Since the issues debated in these disputes are beyond our grasp, one should be willing to admit that neither side can conclusively prove the truth of its position, can dismiss all the objections to it, and can claim that the opponents are willfully in error.[89] Therefore, their good faith must be presumed, and

---

[85] DHC, Manichéens, D: "If anyone were to tell us, with a huge apparatus of reasoning, that it is not possible for moral evil to be introduced into the world by the work of an infinitely good and holy Principle, we would reply that this has nevertheless been done, and consequently that it is very possible. There is nothing more foolish than to reason against facts: the axiom, ab actu ad potentiam valet consequentia, is as clear as this proposition, 2 and 2 make 4." See also DHC, Pauliciens, D: "*Ab actu ad potentiam valet consequentia*, is one of the clearest and most indisputable axioms of all metaphysics (...) *this has happened, therefore it is not repugnant to the holiness and goodness of God.*"

[86] DHC, Pauliciens, E: "(...) the way in which evil was introduced under the empire of a sovereign being infinitely good, infinitely holy, infinitely powerful, is not only inexplicable, but even incomprehensible."

[87] DHC, Pauliciens, M: "The dogma that the Manicheans attack must be considered by the orthodox as a truth of fact, clearly revealed; since we must agree that we do not understand the causes or reasons for it, it is better to acknowledge it from the outset, and stop there, and let the objections of the philosophers run as vain quibbles, and oppose them only with silence and the shield of faith." Bayle gives other, non-religious examples of the limitations of reason, which must accept some truths without understanding them. See Solère 2010, 501–508, 514.

[88] EMT: OD IV, 53: "In his *Dictionary*, Mr. Bayle has established these two articles, one of which is that we must believe that God is not the author of sin, and the other that we must agree that we cannot answer the objections by which the Manichaeans show that our systems give God a conduct that does not accord with common notions of goodness, holiness and justice. The first of these two articles must be deemed evidently established in all the places where Mr. Bayle resorts to the maxim that everything God does is well done, and to Revelation, which teaches us that God has permitted sin, that he condemns it and punishes it. For it follows from this that God is not the author of sin. And this certainty must be enough for us, even though we cannot understand how he does not share in the sin of creatures, nor how men have all the freedom we think necessary to be the author of a punishable thing." RQP: OD III, 770a-b: a Supralapsarist "does not know how God's goodness agrees with the decree of sin, but *Scripture convinces him of this decree, and Reason moreover convinces him that malice, which is an imperfection, cannot be found in God: he therefore does not conclude without the approval of Reason* that there really is agreement between this decree and the sovereign goodness of God." See Solère 2015, 423–424.

[89] DHC, Synergistes, B: It is precisely on matters where both sides face similar problems that "we should most promptly practice a mutual tolerance." Here again, I am in full agreement with Levitin

the conviction they uphold in conscience must be respected.[90] This is the third front (in addition to the Catholic and anti-Calvinist polemics) on which Bayle has to fight in the *Dictionary*: religious tolerance. While Bayle is in fact in doctrinal agreement with Jurieu on the tenets of Calvinist theology, their relationship broke down when Bayle published the *Commentaire Philosophique*, in which he argued for unrestricted toleration on the basis of the rights of erroneous conscience, a view that Jurieu attacked in *Des droits des deux souverains en matière de religion*, 1687 (Levitin 2022, 241 and 367). The "Manichean" articles cleverly use Jurieu's own reflections on the intractability of the question of predestination to remind him that one can invincibly wrong in such matters.

7.9 Conclusion

Let us conclude. After having examined Bayle's objections to traditional Christian responses to the problem of the existence of evil in a world created and governed by a good God, we have wondered why Bayle felt the need to conjure the ghosts of Zoroaster or Mani as spokesmen to formulate these objections, given that he himself points out the flaws and implausibility of their conceptions, and could have found a more robust diarchist system in the works of Plutarch, whom he knew very well and does quote in the course of the same discussions. Looking at the wider context, we discovered that the accusation of "Manicheism" was frequently used in religious controversies in Bayle's day, especially by the Catholics and the Arminians to attack the Calvinist doctrine of predestination. This is why Bayle in his turn uses the specter of Manicheism: he wants to give these critics a taste of their own medicine. In lieu of the narrative according to which Bayle employs Manichean strawmen to secretly undermine Christianity, while astutely mimicking Jurieu's orthodox views to hide behind his authority (Mori 2003, 409), it has become clear that the message that Bayle sends is that the adversaries of Calvinist orthodoxy had better not accuse it of making God the author of sin, for a diarchist can easily prove that they stumble over the very same objection, despite their anti-predestinationist recourse to human free choice. The charge of Manicheism can be turned around and thrown back at them. Consequently, it is in everyone's interest not to raise the issue of God's responsibility regarding evil, but to let it lie dormant. In this way one will avoid the useless controversies that have ravaged Christianity, and the recognition of the insolvability of this problem will lead to tolerance for all men of good will.

2022, 240–241: "(…) there is a frequent tendency to claim that Bayle moved from the 'rationalism' of the *Commentaire philosophique* to the 'fideism' or 'scepticism' of the *Dictionnaire*, and, in some quarters, to claim that this shift betrays his theological insincerity. Nothing could be further from the truth. What has been called 'fideism' was simply Bayle's insistence (…) that the dogmatic truths of religion, especially concerning the mysteries, were incapable of philosophical confirmation and that any ignorance or error concerning them was inevitably 'invincible'."

[90] On the foundations of Bayle's theory of tolerance, see Solère 2016; García-Alonso 2021.

Bibliography

Primary Literature

Adam, Jean. 1668. *Réponse à la lettre de M. Daillé*. Poiters: Jean Fleuriau.
Alexander of Lycopolis. 1974. In *An Alexandrian Platonist against dualism. Alexander of Lycopolis' treatise "Critique of the doctrines of Manichaeus"*, ed. P.W. Van der Horst and J. Mansfeld. Leiden: E.J. Brill.
Augustine. 1887. *Contra Faustum*. In *Nicene and post-Nicene fathers, first series, St. Augustine: The writings against the Manichaeans, and against the Donatists*, ed. Philip Schaff, vol. 4. Buffalo, NY: Christian Literature Publishing.
Averroès. 1996. *Discours décisif*, ed. Marc Geoffroy, introd. Alain de Libera. Paris: Flammarion.
Bayle, Pierre. 1740. *Dictionnaire historique et critique* [= DHC], 5ᵉ éd., 4 vols. Amsterdam/Leiden/The Hague/Utrecht: P. Brunel et al.
———. *Œuvres diverses de Mr Pierre Bayle, professeur en philosophie et en histoire à Rotterdam* [= OD], 4 vols. The Hague: P. Husson et al., 1727–1731.
Daillé, Jean. 1669. *Réplique de Jean Daillé aux deux livres que Messieurs Adam et Cottiby ont publié contre lui*. 2nd ed. Genève: Jean-Antoine & Samuel de Tournes.
De La Valette. 1686. *Parallèle de l'hérésie des Albigeois et de celle du calvinisme, dans lequel on fait voir que Louis le Grand n'a rien fait qui n'eût été pratiqué par S. Louis*. Paris: Lambert Roulland.
Erasmus, Desiderius. 1935. In *De libero arbitrio*, ed. J. von Walter. Leipzig: A. Deichert.
Fisher, John. 1597. *Assertionis lutheranae confutatio*. Wurzburg: Georg Fleishmann.
Jansenius, Cornelius. 1640. *Augustinus*, ed. Jacob Zegers. Louvain: Zegers.
Laurentius a Brundusio. 1930. *Lutheranismi hypotyposis*. In *Opera Omnia*, vol. II-1. Padua: Officina Typographica Seminarii.
Le Cène, C., and J. Le Clerc. 1685. *Entretiens sur diverses matières de théologie. Première partie*. Amsterdam: Nicolas Schouten.
Pascal, Blaise. 1963. *Œuvres complètes de Blaise Pascal*, ed. Louis Lafuma. Paris: Éditions du Seuil.
Plutarch. 1936. *Moralia, volume V: Isis and Osiris. The E at Delphi. The oracles at Delphi no longer given in verse. The obsolescence of oracles*, transl. Frank Cole Babbitt Cambridge, MA: Harvard University Press.
———. 1976. *On the generation of the soul in the Timaeus*, ed. and transl. Harold Cherniss. Cambridge, MA: Harvard University Press.

Secondary Literature

Beck, Roger. 2002. Zoroaster, V. As perceived by the Greeks. *Encyclopædia Iranica*. www.iranicaonline.org/articles/zoroaster-iv-as-perceived-by-the-greeks. Accessed 2 Apr 2023.
Bidez, Joseph, and Franz Cumont. 1938. *Les mages hellénisés. Zoroastre, Ostanès et Hystaspe d'après la tradition grecque*, 2 vols. Paris: Les Belles Lettres.
Carbonnier, Jean. 1955. De l'idée que le protestantisme s'est faite de ses rapports avec le catharisme, ou des adoptions d'ancêtres en histoire. *Bulletin de la Société de l'Histoire du Protestantisme Français* 101: 77–87.
Dillon, John. 1999. Monotheism in the gnostic tradition. In *Pagan monotheism in late antiquity*, ed. Polymnia Athanassiadi and Michael Frede, 69–79. Oxford: Clarendon Press.
———. 2002. Plutarch and god: Theodicy and cosmogony in the thought of Plutarch. In *Traditions of theology: Proceedings of the VIII symposium Hellenisticum*, ed. Michael Frede and André Laks, 223–238. Leiden: Brill.

Duchesne-Guillemin, Jacques. 1998 (revised 2016, last updated July 3, 2023). Zoroastrianism. In *Encyclopedia Britannica*, online edition. www.britannica.com/topic/Zoroastrianism. Accessed 9 Sept 2023.
Febvre, Lucien. 2003. *Le problème de l'incroyance au XVIe siècle*. Paris: Albin Michel.
García-Alonso, Marta. 2021. Persian theology and the checkmate of Christian theology: Bayle and the problem of evil. In C. Masroori, W. Mannies, J. C. Laursen, 75–100. Liverpool: Voltaire Foundation and Liverpool University Press.
Gnoli, Gherardo. 1985. *De Zoroastre à Mani: Quatre leçons au Collège de France*. Paris: Klinckseck.
———. 1996 (last updated Dec. 1, 2011). Dualism. In *Encyclopædia Iranica*, online edition. www.iranicaonline.org/articles/dualism. Accessed on 5 Feb 2023.
Hickson, Michael. 2013. Theodicy and toleration in Bayle's D*ictionary*. *Journal of the History of Philosophy* 51 (1): 49–73.
———. 2016. Introduction. In Pierre Bayle, *Dialogues of Maximus and Themistius*, ed., transl., and introd. by Michael W. Hickson, 1-101. Leiden-Boston: Brill.
Jossua, Jean-Pierre. 1977. *Bayle ou l'obsession du mal*. Paris: Aubier-Montaigne.
Krumenacker, Yves. 2006. La généalogie imaginaire de la Réforme protestante. *Revue Historique* 308 (2): 259–289.
Labrousse, Elisabeth. 1963. *Pierre Bayle*. Tome I. The Hague: Martinus Nijhoff.
Lange, Armin, et al., eds. 2011. *Light against darkness: Dualism in ancient Mediterranean religion and the contemporary world. (Journal of Ancient Judaism, Supplements II.)*. Göttingen/Oakville, CT: Vandenhoeck & Ruprecht.
Levitin, Dmitri. 2022. *The kingdom of darkness. Bayle, Newton, and the emancipation of the European mind from philosophy*. Cambridge: Cambridge University Press.
Masroori, Cyrus, Whitney Mannies, and John Christian Laursen, eds. 2021. *Persia and the Enlightenment*. Liverpool: Voltaire Foundation and Liverpool University Press.
Masroori, Cyrus, and Laursen, John Christian. 2021. The background: European knowledge of Persia before the Enlightenment. In C. Masroori, W. Mannies, J. C. Laursen, 17–42.
Miller, Patrick Lee. 2011. Greek philosophical dualism. In A. Lange et al., 107–144.
Mori, Gianluca. 2003. Pierre Bayle on scepticism and "common notions". In *The return of scepticism, from Hobbes and Descartes to Bayle*, ed. Gianni Paganini, 393–413. Amsterdam: Kluwer.
Popkin, Richard. 1967. Manicheanism in the Enlightenment. In *The critical spirit. Essays in honor of Herbert Marcuse*, ed. Kurt H. Wolff and Barrington Moore Jr., 31–54. Boston: Beacon Press.
Puech, Charles-Henri. 1979. *Sur le Manichéisme et autres essais*. Paris: Flammarion.
Skjærvø, Prods Oktor. 2011. Zoroastrian dualism. In A. Lange et al., 55–91.
Solère, Jean-Luc. 2010. Scepticisme, métaphysique et morale: le cas Bayle. In *Les "Éclaircissements" de Bayle*, ed. H. Bost and A. McKenna, 499–524. Paris: Honoré Champion.
———. 2015. Création continuelle, concours divin et théodicée dans le débat Bayle-Jaquelot-Leibniz. In *Leibniz et Bayle: confrontation et dialogue*, ed. C. Leduc, P. Rateau, and J.-L. Solère, 395–424. Stuttgart: Franz Steiner Verlag.
———. 2016. The coherence of Bayle's theory of toleration. *Journal of the History of Philosophy* 54 (1): 21–46.
———. 2023a. La théodicée de Pierre Bayle. In *Dieu d'Abraham, Dieu des philosophes: Révélation et rationalité*, ed. O. Boulnois, 171–193. Paris: Vrin.
———. 2023b. Bayle, les notions communes, et les silences de la raison. In *L'esprit critique dans l'Antiquité. La naissance de la théologie comme « science »*, ed. Oliver Boulnois, Ph. Hoffmann, Cl. Lafleur, and J.-M. Narbonne, 459–491. Paris: Les Belles Lettres.
———. 2024. Bayle, l'"averroïsme," et le discours théologique. In *Connaître Dieu. Métamorphoses de la théologie comme science dans les religions monothéistes*, ed. O. Boulnois, S. De Franceschi. and Ph. Hoffmann, 575-595. Leiden: Brill.
Stausberg, Michael. 1998. *Faszination Zarathushtra: Zoroaster und die europäische Religionsgeschichte der frühen Neuzeit*. Berlin/New York: Walter de Gruyter.

———. 2000. Pierre Bayle (1647–1706) und die Erfindung des europaischen Neomanichaismus. In *Studia Manichaica: IV. Internationaler Kongress zum Manichäismus, Berlin, 14.-18. juli 1997*, ed. Ronald E. Emmerick, Werner Sundermann, and Peter Zieme. Berlin: Akademie Verlag.

———. 2005. Zoroaster, VI. As perceived in Western Europe. *Encyclopaedia Iranica, online edition*. www.iranicaonline.org/articles/zoroaster-perceived-in-europe. Accessed Apr 2023.

Sundermann, Werner. 2009. Manicheism, I. General survey. *Encyclopædia Iranica,* online edition. www.iranicaonline.org/articles/manicheism-1-general-survey. Accessed 2 March 2023.

Tardieu, Michel. 2008. *Manicheism*. Urbana/Chicago: University of Illinois Press.

Thévenaz, Pierre. 1938. *L'âme du monde, le devenir et la matière chez Plutarque*. Neuchâtel: Paul Attinger.

Van der Lugt, Mara. 2016. *Bayle, Jurieu, and the Dictionnaire historique et critique*. Oxford: Oxford University Press.

Walther, David. 1968. Were the Albigenses and Waldenses forerunners of the Reformation? *Andrews University Seminar Studies* 6 (2): 178–202.

Chapter 8
Bayle and the American and African Atheists

John Christian Laursen

Abstract Pierre Bayle did not have much to say about the religions of the native Americans or non-Christian Africans. When he did refer to them, it was usually to carry on by other means his critiques of Christianity. Atheist Americans and Africans are cited to demonstrate that atheists can have equal or better morals and politics than most Christians. From early work in 1682 to posthumous work published in 1707 Bayle returns to variations on this theme. One purpose is to justify toleration of atheists (and of other religions): they are harmless. Another point is to bring out the hypocrisy of Christians in claiming higher morals than others, but then not living up to them. That is part of a more general point that people everywhere do not live by their professed ideals but by their temperament, pleasures, ambitions, and habits. Along the way Bayle considers the charges of atheism against Spinoza, and while rejecting elements of Spinoza's philosophy he vindicates him from charges of immorality on account of his atheism. The upshot of all of this is a clear case for freedom of religion and freedom of thought, and a powerful rejection of persecution on the basis of people's beliefs.

Keywords America/Americans · Africa/Africans · Atheism · Pierre Bayle · Catholics · Virtuous atheists

---

This paper was made possible by the research project *Contra la ignorancia y la superstición: las propuestas ilustradas de Bayle y Feijoo* (PID2019-104254GB-100) financed by the Spanish Ministry of Education and Science.

J. C. Laursen (✉)
Department of Political Science, University of California, Riverside, CA, USA
e-mail: johnl@ucr.edu

© The Author(s), under exclusive license to Springer Nature Switzerland AG 2024
M. García-Alonso, J. C. Laursen (eds.), *The Importance of Non-Christian Religions in the Philosophy of Pierre Bayle*, International Archives of the History of Ideas Archives internationales d'histoire des idées 251, https://doi.org/10.1007/978-3-031-64865-6_8

Pierre Bayle does not seem to have been a serious student of the religions of North and South America or of Africa. Rather, he was interested in them for the sake of his polemics against prevailing conventional wisdom. One of the most important of these polemics was the question of atheism. It was widely believed that atheism was both a major challenge to Christianity and a terrible danger to society. Bayle fought that view from his early *Various Thoughts on the Appearance of a Comet* (1682) to his last works from the early 1700's. It seems clear that his references to the culture and religion of the Native Americans and Africans were usually designed to contribute to that polemic.

8.1 Overview of Bayle on Atheism

There should be no doubt that the question of atheism was high on the list of early modern debate and action. It has sometimes been said that atheism was not even conceivable in the period, as a "question mal posée", but at least one large philosophical work of atheism had been penned by 1659, the *Theophrastus redivivus* (Canziani and Paganini 1981), which has recently been attributed to Guy Patin (Mori 2021). The philosopher Benedict de Spinoza, writing in the 1660s and 1670s, was widely accused of atheism, and Bayle added to that accusation, as we shall see below. Bayle's contributions to the debate were enormously influential and it has been said that he was "the author with the most impact on the atheism of the Enlightenment" (Zorrilla 2022, 136). His reflections on atheism can still inspire reflection in our time (Kelly 1988).

In his first major published work, *Various Thoughts on the Occasion of a Comet* (1683), Bayle broached the question of the possibility of virtuous atheists. The conventional wisdom was that atheists, with no fear of punishment by God, would break the law whenever they could get away with it. Human life would become a chaotic anarchy of violence. The strong would oppress the weak and the rich would oppress the poor. Fear of God's justice and punishment after death were what kept society and governments honest and stable. In *Various Thoughts* Bayle dismantled many aspects of this wisdom.

In Bayle's account, whether or not one is an atheist makes little difference in one's behavior. Most of the time we do not live by our principles or stated beliefs, but according to our "temperament, the natural inclination toward pleasure, the taste one contracts for certain objects, the desire to please someone, a habit gained in the commerce of one's friends" (PD: OD III, 88; Bayle 2000, 169). It is factors such as temperament and taste, not opinions about God and religion, that determine what one does (PD: OD III, 92–94; Bayle 2000, 178–180). Passions, pride, and the "mechanical constitution of their nature" are the major factors, not philosophical or theological ideas (PD: OD III, 109; Bayle 2000, 210–211). That means that rewards and punishments, and appeals to pride, reputation, and vanity are enough to keep social peace, with no necessary reference to God (PD: OD III, 109–110, 115; Bayle 2000, 212–213, 223).

Bayle pointed out that history tells of atheists who were good people (PD: OD III, 111; Bayle 2000, 214). We also have records of religious people who were just as bad as what people say about atheists: people at the court of Catherine de Medici, "although they believed in God,[...] were capable of every sort of wickedness" (PD: OD III, 100; Bayle 2000, 193). So atheism and religion have no necessary connection to social peace and political stability. There is much more on atheism in the *Various Thoughts on the Occasion of a Comet*.

In many of his later works Bayle pushed variations on the theme of the virtuous atheist and the possibility of peace and prosperity in an atheist society. In his *magnum opus*, the *Historical and Critical Dictionary* of 1697 (and revised and expanded many times), there is a great deal more about atheists, and naturally he was accused of atheism for that. He added a "Clarification concerning atheists" to the second edition of that work (1702), in which he did not retract his views, but rather reaffirmed that atheism could be harmless to society. In the absence of moral truths based on truths about God, people might nevertheless behave well because of their "temperament, education, liveliness of ideas of virtue, love of glory, or dread of dishonor", he wrote (DHC, "I. Eclaircissement"; Bayle 1991, 407).

By the time of his last works Bayle had said a great deal about atheism. In many works, even as late as his posthumously published *Rèponse aux questions d'un provincial* of 1706, Bayle brought out the case of the Chinese, who, in parallel to what he said about the Americans, combined what Bayle interpreted as atheism with good morals and stable politics (RQP: OD III, 958, 966, 983, 988; see also CPC: OD III, 227–229, 397; also see the chapter by Marta García Alonso in this volume). There has been much debate about whether or not Bayle himself was an atheist (Popkin 2003; Bost 2006; Mori 1999; McKenna 2015), but there is not much debate about the fact that he defended the possibility of a virtuous atheist society.

There is one glaring anomaly in Bayle's discussions of atheists. In the *Philosophical Commentary* of 1686 he wrote that the magistrate "may and ought to punish those who sap or weaken the fundamental Laws of the State; and of this number we commonly reckon those who destroy the Belief of a Providence, and the fear of divine Justice" (CP: OD II, 431; Bayle 2005, 243). In addition, an atheist "can never plead that saying of St. Peter, *It is better to obey God than Men*" (CP: OD II, 431; Bayle 2005, 243). It is worth noting that this endorsement of the position that magistrates can justifiably persecute atheists is not a ringing endorsement. It is undermined by the qualification "we commonly [*on a coûtume*] reckon", especially coming from an author who dedicates much of his work to refuting common reckonings. And atheists could answer that they do not wish to plead with St Peter, but would rather say that *It is better to obey the laws than men*. Nevertheless, read alone, this passage would align Bayle with the anti-atheists, yet we know that elsewhere he had rejected the arguments for the nefarious consequences of atheism. The most plausible hypothesis for explaining this discrepancy is that he was writing the *Philosophical Commentary* to cry out against both the repression of the Protestants in France by the Catholics and Protestant intolerance where they had the power, and he did not want to muddy the waters or distract his readers from that issue by raising the issue of virtuous atheists. He could do this without too much hypocrisy because many of his writings were written in other voices than his own: in *Philosophical Commentary* the author of the text is said to be an Englishman (CP: OD II, 355; Bayle 2005, 5). Since most English people followed Milton and Locke in believing that atheists should not be tolerated, expression of this opinion could add credibility to the idea that the book was written by an Englishman. The Englishman does not speak directly for Bayle, so Bayle cannot be accused of inconsistency or hypocrisy.

There were other good reasons for Bayle to downplay his defenses of atheism from time to time. People of the time were extraordinarily sensitive to atheism, suspecting it everywhere and punishing it harshly. Bayle had published the *Various Thoughts* anonymously, surely out of fear of persecution for alleged atheism, but he was suspected of having written it. He lost his job as a teacher at the *Ecole Ilustre* in 1693 in part because of charges of atheism (Bost and McKenna 2006). So he might well have put the anti-atheist words in the mouth of the Englishman in order to throw off suspicions that he was the actual author. In any case, anti-atheism was the conventional wisdom so it would not have been a surprise to find it in an author who was arguing for an end to Catholic persecution of the Protestants. It would be understood that this was one thing both parties agreed upon.

8.2 Americans in the *Historical and Critical Dictionary* and Other Writings

Word searches in Bayle's letters do not turn up any references to the Americans, and *Nouveaux de la Republique des Lettres* contains only one: a reference in a letter from a reader to a book about the survival of Canadian natives for weeks on water alone (NRL: OD I, 229). Considering the vast size and wide-ranging themes of Bayle's *Historical and Critical Dictionary*, it is somewhat surprising that he does not say much about the Americans in it. Numerous word searches have found only the following brief mentions of aspects of their life and culture, although there may be a few more. The same goes for other writings which will be reviewed below, except for one of his later books, which will be discussed in the next section. The Americans do not appear to be a major focus of his interest.

Charnley 1990 attempted to catalogue all of Bayle's references to the Americas in the *Dictionary*, and there are not many. In volume 1 he reports on the book by Christophe de Acuña, a Jesuit who traveled down the Amazon from Peru to Para, but there is nothing about religion in his report (DHC, Acuña, Christophe de). In volume 2 the entry on "Leon (Pierre Cieça de)" cites the reports of this author of a *History of Peru* of 1553 to the effect that the Peruvians did not get their vices from the Christians, but already had them before the Christians came. They ate their children by enslaved women, they worshipped the sun and offered their teeth to it, in Cartagena they do not marry virgins but expect a young woman to have earned a dowry by prostitution before marriage, they engage in sodomy, and much more described as the corruption of their manners (DHC, Leon (Pierre Cieca de), A).

8 Bayle and the American and African Atheists

There is one other reference to letters from Peru from a Jesuit, with nothing about the content of the letters (DHC, Hay (Jean), A). Volume 3 has a short article on Marc Lescarbot, described as the author of a *History of New France* who lived there for some time. Bayle notes only that he is using the second edition, from Paris in 1611, which began with a short description of the voyage to New France of Jean Verazzan (DHC, Lescarbot (Marc), A). I have found nothing in volume 4 except for the index. If this is all that Bayle makes of the Americans in his *Dictionary*, we can see that the Americas did not make up a significant part of his world view in the period in which he was writing it.

Wide reading and word searches in many of Bayle's other works do not turn up much about America or the Americans, either. In the *Various Thoughts on the Occasion of a Comet*, among all of the discussion of atheism that we have seen above, there are few mentions of America. Earthquakes in Canada and Peru are mentioned, but not to make any point about culture or religion (PD: OD II, 31; Bayle 2000, 60). Human sacrifice in Peru and Mexico is mentioned only to say that the Spanish suppressed it (PD: OD II, 48; Bayle 2000, 91). Charnley 1990 also surveyed the references to the Americas in Bayle's other writings, and they are sparse.

In his *Critique Générale de l'Histoire du Calvinisme* of Louis Maimbourg of 1682, Bayle quotes the Catholic historian to the effect that violence is the mark of a false church and then writes a syllogism: "All religion, according to these people, that troubles a long-established religion by violent measures, is false: the Catholic religion troubled the long-established religion of the Americans by violent means: therefore, according to these people, the Catholic religion is false" (CG: OD II, 37).[1] There is no further analysis of the religion and culture of the Americans: this is merely a critique of the Spaniards. In *Nouvelles Lettres Critiques sur l'Histoire du Calvinisme* of 1685, Bayle observes that the arguments of the Catholics in favor of their religion were "good to tell to children, or to poor Americans whose attention is susceptible to everything we want to tell them", which reveals little or no knowledge of the Americans, and only of a stereotype that they were similar in some ways to children (NLCG: OD II, 239).[2]

However, a few years later he had more to say about the Americans, as we shall see below.

---

[1] Toute Religion, selon ces Messieurs, qui va troubler une longue possession par des manieres violentes, est fausse: La Religion Catholique est allée troubler la longue possession des Américains, par des manieres violentes: Donc, selon ces Messieurs, La Religion Catholique est fausse.

[2] bon à dire a des enfants, ou à pauvres Américains, qui font un table de atente susceptible de tout ce qu'on veut.

8.3 Africans in the *Historical and Critical Dictionary* and Other Writings

It seems safe to say that Africa was not very important to Bayle. Word searches in the letters in his *Oeuvres diverses* turn up no uses of the words "Afrique", "Hottentot", or "Cafre". His book reviews in *Nouveaux de la republique des lettres* (1684–1687) include no reviews of books on Africa, and otherwise mainly references to things like the claim that Mt. Atlas is in Africa (NR1: OD I, 208), a poem about Africa (NRL: OD I, 757), a kingdom of women warriors in Africa (NRL: OD I 341), and music from Fez (NRL: OD I, 637). There is a reference to "strange and unknown gods [Dieux inconnu & etrangers]" in Asia, Europe, and Africa, but nothing specific about religion in Africa (NRL: OD I, 736). There is no mention of the words "Hottentot" or "Cafre" in the NRL. There are a few mentions of Africa in vol. II of the *Oeuvres diverses*, but they refer to the Donatists in Roman North Africa (CP: OD II, 445). Bayle also reminds the reader that an African custom of throwing foreigners into the sea was no worse than what the foreigners had done to travelers in their own country: "it is not fitting for violators of faith and humanity to find it wrong when they are paid in the same coin " (CP: OD II, 488).[3]

Similarly, there are no mentions of the words "Hottentot" or "Cafre" in volume I of the *Dictionaire*. There are references to things like fishing in Libya (DHC: I, Apicius, C) and a suggestion that the Pope retire to Africa (DHC: I, Braunbom, B), along with many mentions of Roman activities in Africa, but the only references to religion are to persecution of Donatists and Pelagians by Augustine and other Romans (DHC: I, Apulee, 272, B, and more references to Romans in Africa at 283, 364, 392, etc.). Volume II discusses Roman North Africa from time to time, but anything about religion is a dispute among Christians, not a non-Christian religion. Volume III mentions the Gymnosophists of Africa, living on a mountain in Ethiopia near the Nile, who form no community of any kind, and therefore cannot help Bayle's case for virtuous atheists (DHC: II, Gymnosophistes). There are a number of other references, again usually to Roman North Africa, but also to other details, but very little or nothing about non-Christian religion. Vol. IV mentions the Nestorian Christian heresy, and suggests that it owes its survival to the tolerance of the Muslims in North Africa, but draws no lessons about atheism (DHC: IV, Nestorius, E). Vol. V also mentions only Roman North Africa. We can conclude that atheism in non-Roman Africa was not yet a topic of interest for Bayle.

---

[3] n'est point aux violateurs de la foi & de l'humanite de trouver mauvais qu'on les paie en meme monnoie.

8.4 Americans in the *Continuation of the Diverse Thoughts on the Comet* and the *Response to a Provincial's Questions*

One long sustained passage of attention to the Americans can be found in Bayle's *Continuation of the Diverse Thoughts on the Comet* of 1704 (Letters 85–88). These letters follow up on mention of American atheism at two places (CPC: OD III, 190, 228), and then a discussion of atheism in Letters 76–84. In Letters 76 and 77 Bayle discusses authors who claim that atheism is not the worst of opinions, citing, among others, a Catholic who says that Calvinism is worse than atheism, a Protestant who says that Socinianism is worse than atheism, and those who say that heresy is worse than atheism (CPC: OD III, 296ff.). One of his authorities here is Marc Lescarbot, the author of the *History of New France* mentioned above (now in the version of 1617), who wrote that the atheists of Canada were better than the idolators of Mexico, Virginia, and Florida because they did not attribute the glory of the real God to another god, and did not worship dead things (CPC: OD III, 302). Lescarbot is also quoted for preferring the Souriquois[4] and Armouchiquois[5] of Canada to the Virginians because they do not engage in any divine worship, where the Virginians adore and revere insensible objects (CPC: OD III, 302).

Letter 78 contributes to Bayle's preference for intellectual freedom by claiming that whether atheism is better or worse than idolatry is one of the matters that people are entitled to judge for themselves (CPC: OD III, 303–4). This reference to idolators places Bayle's use of the Americans in the larger field of his critique of idolatry, especially in the ancient world but also elsewhere (see McKenna 2015, 249ff.). Letter 79 makes the point that even if it were true that the majority thinks that atheists are worse than idolators, that does not make it more likely that they are (CPC: OD III, 304). Most people reason poorly on this issue, and thus we cannot weigh their judgments equally (CPC: OD III, 306). Letter 80 adds that the poor reasoning of most people can be attributed to the fact that they have not examined the whole question or all of the evidence (CPC: OD III, 306). It ends with a comparison between a person who is not at home and a person who is at home but is involved in criminal activities: which is worse, he asks, with obvious analogy to those who think there is no God and those who think there is, but that their God is pernicious to society (CPC: OD III, 306). Atheists are better than those who have a dangerously mistaken idea of God. This is followed up in Letter 81 by the argument that the doctrine that God is the author or creator of sin has worse consequences for human behavior than atheism (CPC: OD III, 307–8). It implies that if gods cause sin and engage in criminal acts, humans are justified in doing so, too.

Letter 82 was titled "General Idea of Pagan Religion", and it brings out all of the injustices and cruelties of the Trojan War, blaming them in part on Greek religion. The Greeks attributed the war to the jealousies and crime among the pagan

---

[4] The Souriquois are the Micmac of what is now eastern Canada, perhaps especially Prince Edward Island. See MacGregor 2021.

[5] Native Americans of the coast of Maine. See Weiser-Alexander 2021.

goddesses, which humans imitated (CPC: OD III, 308). Atheists would not imitate them, at least not on religious grounds. And finally, Letter 83 made the case that paganism was actually an atheism. The crime of the atheists was to not recognize the true God, and the fact that they did not believe in false gods was actually a good thing, Bayle argues (CPC: OD III, 308). That means that there are two kinds of atheists: those that do not believe in either the true or the false gods, and those who only believe in false gods. According to the logicians, the latter are really atheists because they do not believe in the true god, Bayle concludes (CPC: OD III, 309). Whatever one makes of the logic of this argument, we can observe that Bayle is widening the class of atheists to include people that had usually been classified as believers.

At this point, Bayle brings in the American atheists for substantial discussion. Letter 85 is titled "That it has been found that the savages of Canada have no religion". It refutes a letter from Justus Lipsius of 1597 in which Lipsius tries to convert an atheist. Lipsius asserts as an accepted fact that a people without religion has never been found in either the Old or the New World, not even among the cannibals (CPC: OD III, 312). Bayle points out that Lescarbot's *History of New France* contradicts that, as he has already shown in chapter 77 (CPC: OD III, 302, 311). Now he adds Gabriel Sagard's assertion in his *History of Canada* (1636) that although Cicero had long ago asserted that all peoples had some sense of religion, nevertheless "our Hurons, & Canadians seem to have had no practice or exercise of religion that we have been able to discover, indeed although they admit a first principle and creator of all things, and thus a divinity like the rest of the nations, nevertheless they do not pray to anything and live almost like animals without adoration, without religion, and without vain superstition under the shadow of a religion. They do not speak of temples or priests, they do not speak of public prayers or prayers in common, and if they do any such prayers or sacrifices, they are not to the first cause or first principle, but to certain powerful spirits that live in particular places" (CPC: OD, 312).[6] Bayle's observation on this passage is that it begins by contradicting Cicero but ends up proving his case. That is because "a nation that prays and makes sacrifices to spirits is not lacking in religion, although it does not worship the principle of everything in particular" (CPC: OD III, 312).[7] He concludes that it is safer to say that "some peoples of Canada are completely without religion, and some others are not without it" (CPC: OD III, 312).[8]

---

[6] nos Hurons, & Canadiens semblent n'en avoir aucune pratique, ny exercice, que nous ayons pû descouvrir, car encore bien qu'ils advouent un premier principe & Créateur de toutes choses, & par consequent une Divinité avec le reste des Nations, si est ce qu'ils ne le prient d'aucune chose, & vivent presque en bestes, sans adoration, sans Religion & sans vaine superstition sous l'ombre d'icelle. De Temples, ny de Prestres, il ne s'en parle point entr'eux, non plus que d'aucunes prieres publiques, ny communes, & s'ils en out quelqu'unes à faire, ou des Sacrifices, ce n'est pas à cette premiere cause, ou premier principe qu'ils les adressent, mais à de certains esprits puissans qu'ils logent en des lieux particuliers.

[7] une Nation qui fait des prieres & des sacrifices à des esprits n'est point destituée de Religion, encore qu'elle ne rende aucun culte particulier au principe de toutes choses.

[8] quelques peuples de Canada sont absolument sans Religion; quelques autres ne le sont pas.

The other author Bayle has mentioned, Lescarbot, was more judicious, Bayle observes. Lescarbot writes of the Souriquois and their neighbors that "I cannot say anything except that since they are lacking in any recognition of God, they engage in no worship, and they have no divine services, they live in a pitiable ignorance" (CPC: OD III, 312).[9] The "Armouchiquois" of the coast of Maine are another "great people who engage in no worship", he wrote (CPC: OD III, 312).[10] The inhabitants of Virginia, unlike those farther north, "believe in many gods" (CPC: OD III, 312). The original God made more Gods to help him govern the world. These people "believe generally in the immortality of the soul and that after death good people are at rest while bad people suffer", Bayle quotes (CPC: OD III, 312).[11] Since the good were their own people and the bad were their enemies, the best thing to do was to kill their enemies. In passing, Bayle suggests, this understanding of the Gods was no better than atheism as far its consequences were concerned: it justified lots of killing (CPC: OD III, 312). The natives of Florida had no knowledge of God or any religion, Lescarbot observed, except to appeal to the sun for victory in war, and Lescarbot did not want to give such appeals the honor of being a religion (CPC: OD III, 312). As Stefano Brogi put it, Bayle is bringing out the point that what we call religions can justify the violation of natural morality, while atheists are at least capable of acknowledging the natural and immutable foundations of at least some basic ethical principles" (Brogi 2022, 155).

Bayle's point about all of these people is that it is not enough to recognize a first principle and creator in order to avoid being an atheist. After all, he writes, the ancient Greek Strato and the seventeenth century philosopher Spinoza recognized a first principle, but they are considered atheists because on their accounts that first principle acts by emanation and is immanent in the world, is determined by natural necessity and does not determine what happens in nature by its own free will, which further means that it does not understand our prayers and thus our prayers cannot bring about a change in the natural course of things (CPC: OD III, 312; see McKenna 2017 and Zorrilla 2022). These are the elements of Bayle's definition of atheism, and thus the Native Americans who recognize a first principle and a creator may nevertheless be atheists.

Letter 86 was titled "Ridiculous opinions of some savages concerning the nature of God." Bayle returned to the Souriquois and reported that they believed that there is one God who had a son and a mother, and together with the sun they made four, although the God was above the other three (CPC: OD III, 313). Instead of a trinity, they had a quadruple God. The son and the sun were benevolent, but the mother was worthless and even ate people, and the "Father, who is God, is not very good" (CPC: OD III, 313).[12] The Souriquois told a story about the supreme God breaking a clay

---

[9] je ne puis dire sinon qu'ils qu'ils sont destituez de toute connoissance de Dieu, n'ont aucune adoration, & ne font aucun service divin, vivans en une pitoyable ignorance.

[10] grand peuple lesquels aussi n'ont aucune adoration.

[11] generalement croyent l'immortalité de l'ame, & qu'après la mort les gens de bien sont en repos, & les méchans en peine.

[12] Pere, qui est Dieu, n'étoit pas trop bien.

pipe that belonged to a man and giving the man another one, saying that as long as it was held by their "grand Sagamo" they would thrive. When it was lost, famine came. The native informant judged that the supreme God was not very benevolent because he let the people's abundance depend upon such a fragile pipe, and being in a position to alleviate their sufferings he let them suffer more than all of the other nations (CPC: OD III, 313). Bayle himself was to say a lot about theodicy, or why God lets people suffer, in several of his works, later to be answered by Leibniz in his *Theodicy* (Leibniz 1710).

Sagard reported that the Hurons believed that the Creator who made the world had a grandmother, and when they were told that it was a contradiction to say that the Creator who has existed for all time has a grandmother, they had no reply (CPC: OD III, 313). They said their beliefs were handed down from father to son, and their narrative came from a figure whose footsteps could still be seen in a certain rock near the river (CPC: OD III, 313). This sort of justification of a belief would remind anyone familiar with other religions of the weakness of some of the proofs of those religions. Further, the grandmother of their God sowed wheat, worked, drank, ate, slept, thus showing that she lived just like the Hurons did (CPC: OD III, 313). Unlike the Creator, who was good, she was evil, and ruined many of the good things that her grandson did (CPC: OD III, 313). Other doctrines that could remind of Christianity were that the animals of the earth belonged to them, that the grandmother lives in the heavens, where there are inhabitants just like on earth, and that she governs the world together with her grandson (CPC: OD III, 313).

When asked if the Hurons worship or pray to their supreme God, they answered that they could do nothing; that he lived too far away to hear their prayers (CPC: OD III, 313). Why do you pray, then?, they were asked, and answered that they prayed and offered presents to spirits that lived in the rivers and rocks, which had immortal souls just like animals and men (CPC: OD III, 313). They added that even though people could not see their actions, nevertheless they could have a beneficial effect (CPC: OD III, 313). Sagard concluded that although they believed in the immortality of the soul, they did not distinguish between the good and the bad, glory or punishment, and expected to be accompanied by their domesticated dogs in the next life (CPC: OD III, 313). They also had preachers who would preach to the fish to make them more likely to swim into their nets (CPC: OD III, 314). The Christian idea of prayer to a supreme God was absent.

In Letter 87 Bayle asserted that what he had reported about the Canadians (Souriquois and Hurons) confirmed what he had said in Letters 83 and 84, summarized above. There were two sorts of Canadians, those who had no religion and those who were really atheists because they had no idea about the true God and were merely superstitious (CPC: OD III, 314). Why were the latter any better than the former?, he asked. Isn't it worse to speak badly about God than to say nothing at all? (CPC: OD III, 314). The true religion would teach that moral virtue was the most important characteristic of God, and the natives who had a religion did not make any connections between their Creator and virtue (CPC: OD III, 314). Bayle's most important point is that he sees no evidence that the natives who have no religion behave any worse than the ones who have a religion, however contradictory and misguided (CPC: OD III, 314). As a comparison, Bayle asserts that good Christians who have the right understanding about God nevertheless cannot resist temptations or control their tumultuous passions and must be constantly reminded about the greatness of God, and they still might not behave well (CPC: OD III, 315). So it could not be expected that a religion of superstitions could make people good. And Bayle adds a comparison to ancient Roman paganism and its inability to make its believers behave well (CPC: OD III, 315).

In concluding his remarks on the Canadians in Letter 87, Bayle observes that "If you make a big case that the Canadians believed in the immortality of the soul, you will be paid in very bad coin. Haven't they given the same honor to the souls of beasts, and the souls they attribute ridiculously to their fish and their axes? Are these derived from the true principles of the dogma of the immortality of the soul? Do they serve in any way to promote good morals? That is the main point" (CPC: OD III, 315).[13] Bayle's big point was that a doctrine such as immortality of the soul does not necessarily lead to better behavior. Thus, atheists of either type, those who have no such beliefs and those who have erroneous beliefs about the soul, may behave equally well or equally poorly. The Americans provided grist for Bayle's mill on the question of the moral effects of atheism.

In Letter 88, Bayle quotes Louis Thomassin to the effect that if any nations have been found that were without any knowledge of God, they were the most savage of nations, almost completely brutish, like the Cafres (of Africa) and Iroquois (CPC: OD III, 316). Bayle's first remark on that is that some of the ancient Greek philosophers were both atheists and among the most polished and learned of all nations (CPC: OD III, 316). His second remark was that the Cafres and Iroquois were no more ignorant or barbaric than many of the other peoples discovered in recent years in Africa and America. (We shall return to this discussion of Africans below.) Neither of the idolatrous or unbelieving Canadians were more polished than the others (CPC: OD III, 316). Bayle quotes Arnauld to the effect that although Mexico and Peru had some form of government, the rest of the natives of the continent lived in sovereign and independent family groups (CPC: OD III, 316). There are differences between the industry, brutishness, and morals of the barbarous peoples, but if anyone thinks that anyone who adores a rock, a piece of wood, or a stream "necessarily has more intelligence and more reason than a man who worships nothing", they are wrong, he writes (CPC: OD III, 316).[14] "Is there any evidence for this proposition?", he asks, and the answer is obviously "no" (CPC: OD III, 316).[15] "If one counts the errors of the Canadian idolators and those of the Canadian atheists, the number will

---

[13] Si vous faites un grand cas de ce que les Canadois ont crû l'immortalité de l'ame, vous vous païez de fort mauvaise monnoie. N'ont-ils pas fait le meme honneur à l'ame des bêtes, & aux ames qu'ils donnoient ridiculement à leurs filets & à leurs haches? Ont-ils tiré de ses veritable principes le dogme de l'immortalité de notre ame? L'ont-ils fait servir le moins du monde aux bonnes moeurs? C'est-là le grand point.

[14] a plus d'esprit & plus de raison nécessairement qu'un homme qui n'adore rien.

[15] Cette proposition-là a-t-elle aucune evidence?

be smaller for the latter" (CPC: OD III, 316).[16] The point is always that atheists do not fare worse than believers in wrong religions, either in knowledge or behavior.

Michael Hickson has argued that Bayle's sources were wrong here, and that he ignored a better source, Jean de Brébeuf in the *Relations des Jésuites de la Nouvelle-France*.[17] It is hoped that Hickson will one day write something about Bayle's neglect of that source and what it would have taught him. But from what we have seen, Bayle might not have cared a great deal about accuracy: what he was looking for was evidence of peoples anywhere who could be considered atheists. That evidence helped him disprove the conventional wisdom that there were no atheists, and that if there were, they would behave worse than believers.

Bayle returned to the issue in Letter 118, presenting an enthymeme: "Atheistic peoples divided into independent families have maintained themselves since time immemorial without any law, therefore there is a stronger reason to think that they could maintain themselves if they were united under a common master and under a code of laws distributing punishments and rewards" (CPC: OD III, 352).[18] He added: "What can you say against that? Don't you have to renounce your second proposition, that irreligion is incompatible with society?" (CPC: OD III, 352).[19]

The issue evidently stayed with Bayle, because it comes up several times in his last major posthumous work, the *Réponse aux Questions d'un Provincial* (1707). On the issue of atheist peoples, he discusses the charge against him that he relies too much on the accounts of travelers, which may be unreliable. But great theologians have accepted their accounts with respect to America, the orient, and Africa (RPQ: OD III, 694–95). "One can only tell them that if the travelers are suspected of misrepresentation when they mention the atheism of some peoples, they could also be misrepresenting when they attribute knowledge of God to certain savage peoples of Africa and America. Would a traveler who is persuaded that it is necessary or advantageous to orthodoxy that the most ignorant and brutal nations have some idea of God have any scruples about inferring such ideas in his reports of what he found among the atheists? Would he have scruples about altering the truth for the sake of a beneficial lie or a pious fraud?" (RQP: OD III, 695).[20] We have been assured that

---

[16] Si l'on comptoit les erreurs des Canadois Idolatres, & des Canadois Athées, le nombre seroit plus petit du côté de ces derniers.

[17] Michael Hickson, personal communication, 2022. See (Brébeuf 1636).

[18] Des peuples athées divisez en families indépendantes se sont maintenus de tems immemorial dans l'Amérique fans aucune loi, Donc à plus forte raison ils se seroient maintenus s'ils se suffent réunis sous un commun maitre & sous un code distributeur de peines & de récompenses.

[19] pouvez-vous dire contre cela? Ne faut-il pas que vous renonciez a votre seconde prétention, qui eft que l'irréligion eft incompatible avec les Societez?

[20] On peut seulement lui représenter que si elles sont suspectes de fausseté entant qu'elles font mention de l'Athéisme de quelques peoples, ells peuvent aussi l'être entant qu'elles attribuente la conoissance de Dieu à certains peoples sauvages de l'Afrique & de l'Amérique. Un voyageur qui sera persuade qu'il est nécessaire ou avantageux à l'orthodoxie, que les nations les plus ignorantes, & les plus brutales aient conservé l'idée de Dieu, ne se fera-t-il pas un scrupule d'inférer dans ses relations qu'il a trouvé des peuples athées? Se fera-t-il un scrupule d'altérer la verité par un mensonge officieux, ou par un fraud pieuse?

peoples without a belief in God believe in the immortality of the soul, the return of spirits, and the augurs, he notes. The conclusion is that those who support the argument from general consent that there is a God and distrust the travelers' accounts when they claim to find atheists should also distrust the travelers when they claim to find that everyone is religious (RQP: OD III, 695). One of those who retails such a beneficial lie is Bayle's nemesis, Pierre Jurieu.[21] He wrote that "the Americans and the cannibals who have no idea about divinity nevertheless believe that after death there are goods and evils, suffering and recompenses" (RQP: OD III, 695).[22] The paradox that Jurieu must not be able to see is that such atheists would have the same reasons for behaving well in this life as the believers.

Bayle approaches religion from many angles in this book, with some evidence from America. He is impressed that "The credulity of the Americans is such that Fernando de Soto, Governor of Cuba and general of Florida persuaded them that Christians are immortal… [and that] nothing of the natives' most secret plans is hidden from them… On this foundation they could not even think about rebelling" (RQP: OD III, 701).[23] In answer to the rationalist theologian Jacquelot, who asks "Why are the Americans savages?", he answers that "If the Americans are content with their condition they will be happy without any knowledge of the fine arts" (RQP: OD III, 831).[24] He cites a report from Virginia that "there are Americans that pay no worship to God because they believe that he does good by necessity and without a choice, and that he does not meddle with human things, so they worship the Devil because they think that he will take away all the goods of God and make them miserable if they do not appease him with their sacrifices and homage" (RQP: OD III, 949).[25] Bayle's comment is that their religion makes them prefer crime to honesty, and thus it makes them worse than someone with no religion (RQP: OD III, 949). The general point here, that "indigenous societies of unbelievers… are more immune to error than indigenous societies of believers" on some issues, has been taken to demonstrate that Bayle thought that primitive atheists were "immune to error", but that is clearly taking the case too far (Bayerl 2007, 16). Relative immunity is not the same as immunity.

---

[21] See Laursen 2018.

[22] les Américains & les Canibales qui n'ont point de sentiment de divinité, croient pourtant qu'aprés la mort il y a des biens & des maux, des peines & des recompenses.

[23] La credulité des Américains a été telle que Fernand De Soto Gouverneur de Cuba & Général de Floride leur persuada que (b) les Chretiens étoient immortels… [and that] rien ne luy estoit caché de leurs desseins les plus secrets… Sur ce fondement ils n'osoient pas même penser à la rebellion.

[24] Pourquoi les Américains sont-ils sauvages?… Si les Américains étoient contens de leurs condition, leur bonheur sufiroit sans aucune connoissance des beaux arts.

[25] il y a des Américains qui ne rendent aucun culte à Dieu, (i) parce qu'ils croient qu'il fait du bien nécessairment, & sans aucun choix, & qu'il ne se mêle point des choses humaines; mais qu'ils adorent le Diable, parce qu'ils croient (k) qu'il leur raviroit tous les biens de Dieu, & les rendroit malheureux, s'ils ne l'apaisoient par leurs sacrifices, & par leurs hommages.

8.5 Africans in the *Continuation of the Diverse Thoughts on the Comet* and the *Response to a Provincial's Questions*

In our discussion of Bayle on the Americans above, we have seen that he occasionally compared them to Africans. In Letter 88 of the *Continuation* he wrote that "One finds in America and in Africa many idolatrous nations as brutish and savage as these two. One is convinced with respect to Africa if one compares together what Mr. Dapper reports about the mores of the inhabitants of the Cafrerie and the mores of the inhabitants of the Negritie" (CPC: OD III, 316).[26] He adds that "we have found atheists in the Mariana Islands where you will find more intelligence and less grossness than in many of the idolatrous peoples of Africa and America" (CPD: OD III, 316).[27] The point is the same as what he says at length about the Americans: atheists do not have to be brutish and savage. Bayle's source here is Dapper, *Description de l'Afrique*, printed in Amsterdam in 1696 (CPC: OD III, 316). Here we see that he is reading recent literature about Africa to complement the older literature about America.

Bayle returned to Dapper in Letter 104, comparing Jurieu's description of Spinoza's idea of God with the Africans. "One does not find that the Hottentots have any divinity that is the object of their adoration. They recognize that there is a sovereign Being, which they call Humma, who makes it rain and blows the winds and gives the heat and cold. But they do not believe that they are obliged to render homage to him" (CPC: OD III, 329).[28] In Letter 118 he reported references from Sallust about Africans who had no civil government but lived peacefully enough together, and argued that the same could be said for atheists (CPC: OD III, 352). He asked the reader to "cast your eyes on the description of the country of the Cafres. You will see that they are atheists, that they are divided into several societies, each under a single chief, that they have laws, and that violators of the laws are punished severely" (CPC: OD III, 353).[29]

Letter 145 was titled "Continuation of the same subject. Examples taken from China and the barbarians of Africa".[30] After some quotes about the very literate

---

[26] On a trouve dans l'Amérique & dans l'Afrique plufieurs Nations idolatres aussi abruties & aussi fauvages que ces deux là. On peut s'en convaincre quant à l'Afrique fi l'on veut comparer enfemble ce que le Sieur Dapper (b) raporte sur les moeurs des habitans de la Cafrerie, & fur les moeurs des habitans de la Nigritie.

[27] on a trouvés Athées dans les iles Marianes vous leur trouverez plus d'esprit ou moins de grossiéreté qu'il n'y en a dans plusieurs peuples idolâtres de l'Afrique & de l'Amérique.

[28] On ne remarque point que les Hottentots aient quelque Divinité, qui soit l'objet de leur adoration. Ils reconnoissènt bien qu'il y a un Etre fouverain, auquel ils donnent le nom de Humma, qui fait tomber la pluye & souffler les vents, & qui donne le chaud & le froid. Mais ils ne croyent pas qu'on soit obligé de luy rendre hommage.

[29] jetter les yeux fur la description du Païs des Cafres. Vous y trouverez qu'ils sont Athées, qu'ils sont divisez en quelques Sociétez, chacune sous un seul chef, qu'ils ont des loix & qu'ils en punissent févérement les infracteurs.

[30] Continuation du meme sujet. Exemples tirez de la Chine & des barbares de l'Afrique.

Chinese, he turns to Africa for evidence that unlettered peoples could also be atheists. Quoting from Dapper, he wrote that "all of the Cafres are people plunged into the darkest ignorance", but after interactions with the Dutch they began to learn Dutch and "one sees good evidence of natural enlightenment in them" (CPC: OD III, 398).[31] Dapper concluded that "the most barbarian Hottentots do not act without reason, and they know the laws of peoples and of nature" (CPC: OD III, 398).[32] He added that "I should add that their mutual love, their faithfulness, and their unselfishness should make the Christians feel ashamed" (CPC: OD III, 398).[33] We can conclude from one interchange between the Dutch and the Africans, says Bayle, that "the Christians have no idea of equity, and the atheist Africans understand it completely" (CPC: OD III, 398).[34] And his Dutch author, Dapper, testifying against his own people and in favor of the Africans, notes that they take a variety of circumstances into consideration in punishing crime and have a great love for virtue and natural equity with no reference to a divine being (CPC: OD III, 398).

As in the case of the *Continuation*, there is a mention in the *Reponse aux Questions d'un Provincial* of Africa, together with America, on the issue of atheism, but it is not a major issue in this work (RQP: III 694, 695).

8.6 Conclusion

We have seen that Bayle was not very well informed about the religions of the Americans or the non-Christian Africans. He used very few sources of the many that were available in his time, perhaps because he did not trust them, as indicated in one of his comments above. He did not refer to them much, especially considering the volume of his work as a whole. He used what he knew mostly to make points he made many times in other ways and based on other cultures. Perhaps one explanation is that all that he had from the Americans and Africans were travelers' accounts, nothing in their own voices. This was of course because they were not literate cultures which had left texts that he could peruse. Arnauld had pointed out that "In no part of America have we found knowledge of the art of writing. Therefore, there are no books, no studies, no care to cultivate their intelligence and their reason", which Bayle quoted (CPC: OD III, 316).[35] The same could be said for the Africans. Since

---

[31] tous les Caffres sont des peuples plungez dans la derniere ignorance ... on voit en eux de beaux restes de la lumiere naturelle.

[32] les Hottentots les plus barbares n'agissent pas sans raison, & qu'ils savant le droit des gens & de la nature.

[33] j'ose meme dire que leur amour mutuel, leur fidelite & leur desinteressement doit couvrir les Chretiens de confusion.

[34] les Chretiens n'avaioent nulle idee de l'equite, & que les Athees [Africains] en etoient tout penetres.

[35] On n'a trouvé nulle part dans l'Amérique la connaissance de l'art d'écrire. Et ainsi nuls Livres, nulles études, nul soin de cultiver son esprit & sa raison.

perhaps his strongest identification was as a member of the Republic of Letters, and his life was spent processing texts, it is perhaps not surprising that the Americans and Africans did not mean much to him since he could not read them in their own words, or even in translation. They provided him with evidence for his overall argument about atheism and virtuous atheists, but apparently did not change that argument in any significant way.

Bibliography

Bayle's Works

OD: Most of Bayle's works are cited from *OEuvres diverses de Mr Pierre Bayle, professeur en philosophie et en histoire à Rotterdam*, 4 vols. (The Hague, P. Husson et al., 1727–1731), ed. E. Labrousse (Hildesheim: Olms, 1966), using the initials of the work, followed by the volume in which it is found in the *OEuvres diverses* (OD), and the page. CG: *Critique générale de l'Histoire du calvinisme de M. Maimbourg* (1682); PD: *Pensées diverses écrites à un docteur de Sorbonne, à l'occasion de la comète* (1683); NRL: *Nouvelles de la république des lettres* (1684–1687); NLCG: *Nouvelles lettres de l'auteur de la Critique générale de l'histoire du calvinisme* (1685); CP: *Commentaire philosophique sur ces paroles de Jésus-Christ: contrain-les d'entrer* (1686); Supplément: *Supplément au commentaire philosophique* (1688); APD: *Addition aux Pensées diverses sur les cometes* (1694); RQP: *Réponse aux questions d'un Provincial* (1703–1707); CPC: *Continuation des pensées diverses à l'occasion de la comete* (1704); EMT: *Entretiens de Maxime et de Themiste* (1707).

DHC: Bayle's *Dictionary* is cited from the online version available at Gallica: *Dictionaire historique et critique*, ed. P. Des Maizeux, 5th edition, Basel: Brandmuller, 1738.

Where an English translation is used, it is cited from the following:

Bayle, Pierre. 2000. In *Various thoughts on the occasion of a comet*, ed. R. Bartlett. Albany: State University of New York Press.

———. 2005. In *A philosophical commentary on these words of the gospel, Luke 14:23, "compel them to come in"*, ed. J. Kilcullen and C. Kukathas. Indianapolis, Liberty Fund.

———. 1991. First clarification. In *Diccionaire historique et critique. Historical and critical dictionary*, ed. R. Popkin. Indianapolis: Hackett.

General Bibliography

Bayerl, C. 2007. Primitive atheism and the immunity to error: Pierre Bayle's remarks on indigenous cultures. In *Religion, ethics, and history in the French long seventeenth century. La religion, la morale, et l'histoire à l'âge classique*, ed. William Brooks and Rainer Zaiser, 15–28. Oxford/Berlin: Peter Lang.

Bost, H., and A. McKenna. 2006. *"L'Affaire Bayle". La bataille entre Pierre Bayle and Pierre Jurieu*. Saint-Etienne: Institut Claude Longeon.

Bost, H. 2006. *Pierre Bayle*. Paris: Fayard.
Brébeuf, J. 1636. *Relations des Jesuites*. English translation: "The Mission to the Hurons (1635–1637)". In *First peoples: A documentary survey of American Indian history*, 6th ed, ed. C.G. Calloway, 2019, 111–115. Boston: Bedford. Originals and translations at https://moses.creighton.edu/kripke/jesuitrelations/ (Volumes 10, 11).
Brogi, S. 2022. Immorality and intolerance in the Bible? Natural ethicality and the interpretation of scripture in the writings of Pierre Bayle. In *The philosophers and the Bible*, ed. A. Del Prete, A.L. Schino, and P. Tortaro, 143–157. Brill: Leiden.
Canziani, G., and G. Paganini, eds. 1981. *Theophrastus redivivus*. Firenze: La Nuova Italia.
Charnley, J. J. 1990. *The influence of travel literature on the works of Pierre Bayle with particular reference to the Dictionnaire historique et critique*. Durham theses, Durham University: http://etheses.dur.ac.uk/6574/.
Kelly, G.A. 1988. Bayle's commonwealth of atheists revisited. In *Religion, morality, and the law*, series *Nomos*, vol. 30, 78–109. New York: New York University Press.
Laursen, J.C. 2018. Spinoza et les 'mensonges officieux' dans les manuscrits clandestins. Une question et un programme de recherche. *La Lettre clandestine* 28: 81–98.
Leibniz, G.W. 1710. *Theodicy*. La Salle: Open Court.
Leon, Pierre Cieça de. 1553. *La Crónica del Perú*, Sevilla: Montesdoca. English: *The Travels of Pedro Cieza de Leon, A.D. 1532–1550*, tr. C. Markham. London: Hakluyt Society, 1864.
Lescarbot, Marc. 1617. *Histoire de la Nouvelle-France*. Paris: Millot. English: *The history of New France*, tr. W. Grant. Toronto: Champlain Society, 1917 (reprinted New York: Greenwood, 1968).
MacGregor, J.J. 2021. Souris: Of mice and Mikmaq, June 9, 2021, Mayzil of Prince Edward Island, https://www.mayzil.com/main/the-souriquois?rq=Souris.
McKenna, A. 2015. *Etudes sur Pierre Bayle*. Paris: Champion.
———. 2017. Pierre Bayle, historien de la philosophie: un sondage. *Lexicon philosophicum* 5: 21–59.
Mori, G. 1999. *Bayle philosophe*. Paris: Champion.
———. 2021. A la recherche du Nouveau Théophraste. *La lettre clandestine* 29: 85–143.
Popkin, Richard. 2003. *The history of Scepticism from Savonarola to Bayle*. Oxford University Press.
Sagard, Gabriel. 1636. *Histoire du Canada*. Paris: Sonnius. English: *The long journey to the country of the Hurons*, ed. G. Wrong. Toronto: Champlain Society, 1939 (reprinted New York: Greenwood, 1968).
Weiser-Alexander, K. 2021. Armouchiquois Tribe, Legends of America. https://www.legendsofamerica/armouchiquois-tribe/.
Zorrilla, N.L. 2022. El estratonismo en el materialismo ilustrado: el caso Sade. *Daimon: Revista Internacional de Filosofía* 86: 136.

Index

A
Adam, J., 174
Ambrogio, S., 104, 121
Anderson, S., 45
App, U., 100, 101, 107–110, 113–115, 121
Arminius, J., 175, 181
Arnobius, 167, 168
Assmann, J., 63
Atkinson, G., 43
Augustine of Hippo, 10
Averroes, 6, 61–63, 180

B
Bahr, F., 8, 76, 100–122
Balagna Coustou, J., 43
Bayerl. C., 11, 73, 93, 201
Bayle, P., 1, 4–13, 15–36, 41–66, 72–96, 100–122, 152–184, 190–204
Beck, R., 154
Bespier, H., 45, 62, 63
Bevilacqua, A., 46, 49, 50
Bianchi, L., 16, 91
Bibliander, T., 46, 49
Bobzin, H., 46
Bocking, B., 104–107
Booromeo, E., 43
Borgne, É.L., 43
Bossuet, J.-B., 73, 101, 174
Bost, H., 191, 192
Bost, Hubert, 5, 21, 27, 44, 118
Brébeuf, J.de, 200
Brockliss, L., 130

Brogi, S., 197
Busbecq, O.G.de, 43
Büttgen, P., 65

C
Carvalho, M.S.de, 62
Chardin, J., 44, 45, 53
Charnley, J., 11, 44, 73, 81, 82, 103, 192, 193
Cole, S., 139
Cotton, P., 9, 10, 126–149
Crasset, J., 111

D
D'Herbelot de Molainville, B., 50
Daillé, J., 174, 175
Daireaux, L., 45
Dandini, J., 48
Daniel, N., 42
Daston, Lorraine, 97
De Smet, D., 144
Dew, N., 50
Dillon, J., 170
Dubois, B., 104–106
Dyâb, H., 44

E
Erasmus, D., 119, 173
Erpenius, T., 46, 47, 63
Espagne, M., 65

F
Faustus of Mileve
Fénelon, F., 73, 101, 116
Fernández, J., 107
Fisher, J., 173
Flacius Illyricus, M., 175

G
Gallien, C., 47
García-Alonso, M., 1–13, 26, 30, 34, 72–96, 100, 103, 115, 184
Gay, P., 128
Gernet, J., 72, 76, 78
Girard, A., 44
Glei, R.F., 49
Gnoli, G., 153
Gobillot, G., 64
Goodman, M., 127
Gros, J.-M., 7, 86, 90, 95, 103
Gusdorf, G., 64
Guyon, J.M., 101, 116, 118

H
Hamilton, A., 49
Hanne, O., 42
Hashimoto, K., 77
Heyberger, B., 44, 46
Hickson, M.W., 10, 26, 32, 33, 96, 152, 173, 176, 200
Hyde, T., 156

I
Israel, J., 3, 17, 18, 24, 103, 115

J
Jami Catherine, 97
Jansenius, C., 173, 174
Jaquin, F., 43
Jorink, E., 147
Jossua, J.-P., 157, 160, 168
Jurieu, P., 11, 26, 27, 29, 30, 32, 36, 55–57, 59, 60, 73, 79, 95, 116, 117, 137, 139, 175, 176, 178, 179, 184, 201, 202

K
Kelly, G.A., 190
Khayati, L., 52, 63, 64
Koningsveld, P.S.van, 55
Kow, S., 7, 8, 88

L
La Mothe le Vayer, F.de, 91, 92, 145
Labrousse, E., 5, 9, 17, 18, 25, 26, 44, 119, 121, 168
Lactantius, 83, 139, 164
Lagrée, J., 65
Larzul, S., 49
Laurens, H., 44, 50
Laursen, J.C., 12, 27, 29, 86, 94–96, 103, 190–204
Laven, M., 72, 78
Le Cène, C., 175
Léchot, P.-O., 6, 41–66
Leibniz, G.W., 42, 64, 73, 83, 100, 143, 154, 198
Lennon, T.M., 36, 115, 118, 120
Leon, P.C.de, 192
Lescarbot, M., 193, 195–197
Lestringant, F., 43
Levitin, D., 176, 177, 181, 182, 184
Libera, A.de, 62
Loop, J., 47
Lorenzo di Brindisi

M
MacFarlane, K., 41
MacGregor, J.J., 195
Maia Neto, J.R., 96
Maimbourg, L., 193
Malcolm, N., 42, 43, 46, 53, 64
Mandelbrote, S., 51, 64
Mani, 152–184
Marcion, 163, 166
Mardan-Farrukh, 159
Marenbon, J., 126
Marracci, L., 42, 49, 50, 63
Masahide, B., 105
Matar, N., 42
McKenna, A., 33, 191, 192, 195, 197
Melissus of Samos
Meynard, T., 79
Miller, P.L., 169

Mills, S., 47
Minamiki, G., 74
Molinos, M. de, 101, 116, 118
Moréri, L., 100, 129, 141
Mori, G., 7, 27, 30, 32, 33, 86, 93, 95, 103, 119, 190, 191
Morrow, J.A., 46–47
Mungello, D., 72, 75, 76, 78, 80, 81

N
Nisbet, H.B., 65

O
Ochino, B., 181
Ouardi, H., 57

P
Paganini, G., 4, 7, 92, 93, 96, 190
Pascal, B., 73, 133, 174
Paulicians, 133, 135, 155, 157, 159, 168, 177
Pinot, V., 72, 73, 79, 90
Plutarch of Chaeronea, 168
Popkin, R.H., 5, 15, 17, 22, 174, 191
Possevino, A., 110–115, 121
Postel, G., 43, 75
Prideaux, H., 41, 50, 51, 55, 57, 64

R
Richard, F., 43, 49
Richard, Robert, 69
Rizzi, M., 49
Rycaut, P., 44–46, 61, 63
Ryer, A. du, 49

S
Sagard, G., 196, 198
Said, E.W., 50
Salmon, J.H.M., 128
Schino, A.L., 52
Simon, R., 48, 50, 52, 64
Simplicius of Cilicia, 167
Sionita, G., 46, 58
Skjærvø, P.O., 153, 159
Sola, D., 82
Solère, J.-L., 11, 96, 103, 152–184
Spinoza, B., 5, 6, 9, 24, 25, 33, 35, 42, 61–63, 82, 92, 95, 101, 112–114, 116, 119–122

Stausberg, M., 154, 166, 168
Stroumsa, G., 44
Sutcliffe, A., 5, 15–36, 95

T
Tavernier, J.-P., 44, 45, 102
Taylor, H., 130
Thomson, A., 45
Tinguely, F., 43
Tinsley, B.S., 135
Tolan, J., 42, 64
Toomer, G.J., 47
Torabi, D., 50
Torres, L. de, 107
Tottoli, R., 49

U
Üçerler, A., 108

V
Valignano, A., 105, 106, 108–115, 121
van Boxel, P., 41
Van der Cruysse, D., 44, 53
Van der Lugt, M., 6, 20, 26, 35, 42, 56, 57, 59, 137, 141, 157, 158, 177
van Leeuwen, R., 46, 47
Van Reijen, M., 120
Van Ruisbroeck, J., 116
Varani, G., 42
Vatinel, Denis, 69
Vermeir, K., 143
Vigliano, T., 49
Vrolijk, A., 46, 47

W
Weinberg, J., 41, 46
Weinstein, K.R., 95
Weiser-Alexander, K., 195
Wele, M.K., 49
Weststeijn, T., 82, 91
Whelan, R., 36, 44

Z
Zoli, S., 72, 95
Zoroaster, 152–184
Zorrilla, N.L., 190, 197
Zürcher, E., 77